WITHDRAWN

TIME, SPACE, STARS & MAN
The Story of the Big Bang

2nd Edition

TIME, SPACE, STARS & MAN

2nd Edition

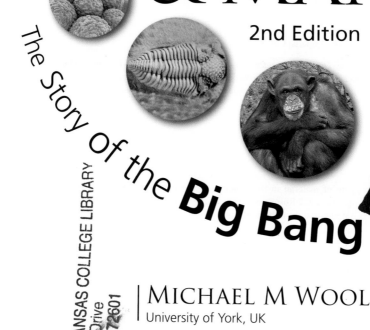

The Story of the **Big Bang**

MICHAEL M WOOLFSON
University of York, UK

Imperial College Press

ICP

Published by

Imperial College Press
57 Shelton Street
Covent Garden
London WC2H 9HE

Distributed by

World Scientific Publishing Co. Pte. Ltd.
5 Toh Tuck Link, Singapore 596224
USA office: 27 Warren Street, Suite 401-402, Hackensack, NJ 07601
UK office: 57 Shelton Street, Covent Garden, London WC2H 9HE

British Library Cataloguing-in-Publication Data
A catalogue record for this book is available from the British Library.

TIME, SPACE, STARS AND MAN
The Story of the Big Bang
2nd Edition

ISBN 978-1-84816-933-3
ISBN 978-1-84816-934-0 (pbk)

Typeset by Stallion Press
Email: enquiries@stallionpress.com

Printed in Singapore by Mainland Press Pte Ltd.

Contents

Introduction

Most scientists work in fields for which the foundations are well established, and I am one of those. I have worked in two quite disparate fields, X-ray crystallography and star and planet formation. That is one of the joys of academic life — if you find something that interests you then you may pursue it, and nobody will stop you from doing so. My two fields of research, together with peripheral material, cover an interesting range. An interest in star formation must include consideration of galaxy structure and, one stage back from there, the beginnings of the Universe that produced those galaxies. A study of the formation of planets must, perforce, include the origin of the Earth and a consideration of the way that it has evolved to its present state. My current interest in crystallography is in the structure of proteins and one stage removed from that is a consideration of the processes that enable life to occur, which depends on the environment offered by the Earth, and also the factors that distinguish living from non-living systems. In 1953, when Crick and Watson produced their revolutionary model of DNA, I worked on the floor above in the Cavendish Laboratory and shared in the excitement of that watershed discovery.

While science has explained a great deal, there are two key events in the history of the Universe for which science has produced no satisfactory explanations: the cause and origin of the Big Bang that originated the Universe and the origin of life that eventually led to the variety of living forms that exist today. If we accept the Big Bang then our scientific knowledge can take us from there to the material structure of the Universe as we know it. If we accept the formation of a first primitive life form then our science can take us through the evolutionary pathways to *homo sapiens* — us.

My life has been spent in education and I strongly believe that a well-rounded, intelligent citizen should understand, amongst other things, the general principles of science that are so important in today's society. I am not saying that all well-rounded citizens must be scientists; that would be absurd and even undesirable. We live in a civilized society and make our contributions in different ways, and all those ways are important. What I began to think about was the possibility of describing all the steps from the Big Bang to the evolution of mankind in words and pictures without resorting to any scientific equations. Could it be done and, if so, could I do it? To make my decision harder I had promised in previous writing *not* to try to explain the Big Bang in future writing — the reason I gave being that I had not got the basic understanding to do so. Ah well — I am human and imperfect!

For most of the time I was at the Cavendish Laboratory, its Head, the Cavendish Professor, was Sir Lawrence Bragg — one of the great British scientists and popular educators of the 20th Century. He made most of his contributions in the form of lectures — and I am going to write — but I thought it worthwhile to read again some of his pronouncements on giving a successful popular lecture. I came across the following passage in his writings:

'A guiding principle of a popular lecture is that of starting with something with which the audience is thoroughly familiar in everyday life, and leading them further with that as a basis. The survey of the new country must be tied on to fixed points which are already in their minds. This is one of the most difficult tasks facing the popular lecturer. He may be honestly trying to avoid technical language; but it goes further than that. He has to put himself in the place of the intelligent layman and realize that ideas and experiences so familiar to him are unexplored country to his listener. This may seem to be stressing the obvious; but I venture to stress it because I have rather special opportunities to assess the effect of popular scientific talks, and they often pass completely over the heads of the audience because an otherwise excellent talk does not establish an initial *rapport* with the listener's knowledge and experience."[a]

Writing the chapters of a book is not the same as giving a series of lectures but there is some commonality in the two processes. A lecturer can engage with his audience on an eye-contact basis and if he has a warm and friendly personality, as Lawrence Bragg certainly had, then his audience will be the willing recipients of his message. A compensating advantage of writing is that the reader can go back and check what he has previously read, something not possible in a lecture.

Another piece of advice that I heard Lawrence Bragg give verbally is never to try to cram too much information into a single lecture. It is common for inexperienced lecturers to do this but after a while they get the message that the problem of preparing a successful lecture is less about what to put in than about what to leave out. Of course that counsel of perfection cannot always be followed in its entirety; there is essential information and somehow or other that must be imparted. What I shall do is to leave out anything that is not *essential* to the task in hand. This will expose me to the risk that scientific purists will criticise what I write as incomplete or even misleading. Angular momentum, a concept I shall be mentioning, is a vector, not a scalar, quantity and I do not mention that. The nature of the quantity is irrelevant in the way that I refer to it so I do not give its nature. Misleading? I think not. If you, as a reader, happen to know the difference between a scalar and vector quantity then that is fine. If you do not, then you certainly do not need to find out what they are for the purpose of understanding what I write here.

I think that the narrative I present gives a logical, sequential and causally-related set of events that go from the Big Bang to man. Others may wish to present a different narrative but the story I present is *my* story as *I* see it.

---

[a] Thomas, J.M. (1991), *Notes Rec. R. Soc. Lond.* 48(2), 243.

Introduction to the Second Edition

After I had completed writing the First Edition of this book and it was well advanced in its production process — in fact, when I was reading the first proofs — I decided that, if I ever had an opportunity to do so, I would change the balance of the presentation. The opportunity arose when I was asked to prepare a Second Edition.

The book has three main themes. The first is the origin and evolution of the Universe up to the stage of producing stars. The second deals with the origin and evolution of the Solar System and, in particular, the early development of the Earth. The third and final theme is concerned with the origin and evolution of life on Earth. The first and last of these themes present fundamental problems for which science has, as yet, no solutions. We do not know the source of the energy that created time, space and all the matter in the Universe, and how this energy manifested itself in the form of the Big Bang. We also have no convincing theory for the way that living matter was produced from non-living matter. The second theme presents no fundamental problem — the matter for producing stars and planets exists all over the galaxy — although how this matter became assembled into planetary systems is a surprisingly difficult problem to solve, one that has exercised the minds of many scientists.

In this edition the parts of the book dealing with the nature and evolution of the Universe and the origin and evolution of the Solar System have been slightly shortened, without the loss of any important material, while that part dealing with evolutionary and biological aspects of life on Earth has been considerably expanded. This modification better represents the relative importance of the three themes and gives a similar level of detail in the presentation of each of them.

Chapter 1

Musing

I am sitting in my study at home, letting my thoughts range freely from one topic to another. Before me there is a wall of books — fiction and non-fiction, scientific and non-scientific — and I try to estimate how many hundreds of years of human effort went into their production. Idly I transfer my gaze to the scene outside the study window and contemplate the bare branches of the trees on the nearby golf course. The New Year has just begun and winter must complete its course before fresh green leaves appear once more. Beyond those trees, some 200 miles away to the south, is London, where my daughter and her family live; they are due to visit next month and I look forward to seeing them all again.

When spring eventually arrives my wife and I will embark on a round-Britain cruise taking in various Scottish Isles, Dublin, the Channel Islands and London. As we get older we have become less adventurous and restrict our travelling to Europe — and the closer parts of Europe at that. In years gone by we travelled much more widely. We went many times, on a biennial basis, to China where I had scientific collaborators who were, and still are, my friends. Our longest journey together was to a conference in Perth, Australia, in 1987. It was an enjoyable visit and put to rest many misconceptions. There was no feeling of isolation there; on the contrary one felt part of a lively and vibrant community. We much admired, and somewhat envied, the quality of life we found in Perth. From Britain one cannot travel much further than that in this finite world of ours. To travel further one must leave the world and that is a privilege of the very few.

Not long ago I listened to a radio discussion about plans for new manned missions to the Moon and the aspirations of the various space agencies, including that of China, to send men to Mars. I wonder how feasible that really is. The discussion was detailed, and involved experts in the field, but one topic that seemed to be absent was that of the safety of the people involved. The Sun is a very active body. In its quieter periods there is a solar flare (Figure 1.1) about once per week. Every eleven years or so it goes through a more violent phase when solar flares tend to be larger and may occur several times a day. Solar flares are violent explosions on the surface of the Sun, releasing large quantities of very hard, penetrating X-rays and energetic charged particles, particularly protons, which are also very penetrating. The most violent solar-flare eruptions, called X-class flares, can have a major effect on terrestrial activities, despite the strong shielding effect of the Earth's atmosphere. In 1989 an X-class flare caused a wide-spread power failure in the Canadian province Quebec and an even stronger flare, on 12th November 2003, disrupted radio communications in California and subjected astronauts, and even some air passengers flying in the stratosphere, to X-ray doses equivalent to that from a medical chest X-ray. Fortunately the main blast from that particular flare was not towards the Earth! Space suits give little protection from the most penetrating solar-flare radiation and spacecraft give partial, but not complete, protection. Scientists working on the International Space Station have been exposed to radiation levels well above average terrestrial levels for long periods without noticeable

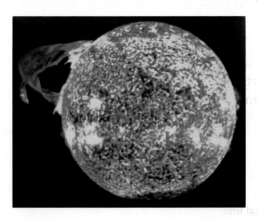

Figure 1.1   A large solar flare (NASA).

harmful effects. Nevertheless, it seems uncertain that astronauts, spending a year or so in space to get to Mars and back, could be adequately protected against unexpected major solar flares, especially if the maximum radiation output was in their direction. There are no such problems with unmanned missions although the working of scientific instruments can be, and has been, affected by radiation from solar flares.

Spacecraft have explored to the outermost reaches of the Solar System; since their launches in 1977 the Voyager I and II spacecraft have left the region of the planets and are at a distance of more than 100 astronomical units from the Sun, where an astronomical unit is the average Sun–Earth distance[a]. The boundary of the Solar System is not something that can be defined with certainty. Beyond Pluto, once considered the furthermost planet but, since 2006, demoted to the status of 'dwarf planet', there exists a swarm of small bodies at least one of which, named Eris, is larger than Pluto. This region, known as the Kuiper Belt, stretches an unknown distance outwards from the Sun. What *is* known is that, orbiting the Sun at distances of tens of thousands of astronomical units, there are comet-like bodies, estimated to number $10^{11,b}$ (one hundred thousand million), in a system known as the Oort Cloud. Once in a while these bodies are gravitationally nudged by passing stars, or other massive astronomical objects, and then some of them are pushed into orbits taking them close to the Sun, when they are observed from Earth in the familiar form of a comet, as seen in Figure 1.2. An average of about one comet per year is produced in this way, although most of them are not very spectacular and can only be seen with telescopes.

If the Oort Cloud is considered to be a part of the Solar System then the system stretches out a large fraction of the way to the nearest star to the Sun, Proxima Centauri. This is at a distance of 270,000 astronomical units. However, when we consider the distances of stars, or entities even further away, the astronomical unit is an inconveniently small unit of distance. In astronomy, and indeed in life in

[a] One astronomical unit is approximately 150 million kilometres or 93 million miles.
[b] 10^{11} represents eleven 10s multiplied together or 100,000,000,000.

Figure 1.2   Comet West (1976) showing twin tails.

general, one always has the problem of comprehending quantities of interest. Most people have a reasonable idea what a kilometre represents in terms of distance, so that when told that York is 320 kilometres from London they can relate to that information. However, although most people also have a reasonable idea what a centimetre is, they would less readily relate to the information that York is 32 million centimetres from London. Similarly we better understand the performance of an athlete when we are told that he has run 100 metres in 10 seconds than if we were told that he had run 10,000 centimetres in $1.1574 \times 10^{-4,c}$ days. To some extent it is a matter of what we are used to, but it also depends on the fact that we have a better feel for the meaning of small numbers than for those that are either very large or very small. In astronomy, when distances of stars or other distant objects are concerned, the light year is a convenient

---

$^c 10^{-4}$ represents $1/10^4$ or 0.0001.

unit. It is the distance that light travels in a year and, since the speed of light is 300,000 kilometres per second and there are 3.156×10^7 seconds in a year, the light year is about 9.5×10^{12} kilometres. On that scale Proxima Centauri is 4.2 light years from the Sun. When we are looking at Proxima Centauri we are seeing it as it was 4.2 years ago. If it were suddenly to explode then we should find out that it had done so 4.2 years after the event.

The Sun is what is known as a *field star*, which is to say that it moves through space without stellar companions. About two thirds of all stars exist in the form of *binaries*, which are pairs of stars that orbit around each other. These binary pairs can also have the property of field stars in that they travel without other companions. However, not all stars are field stars and large numbers of them exist within clusters, of which there are two main kinds. The first of these consists of anything from a hundred to a few thousand stars and these are known as *open clusters* or, sometimes, *galactic clusters*. A very beautiful example, shown in Figure 1.3, is the Pleiades cluster. This is a cluster of about 500 stars of which seven are very bright, and there are biblical references to it. The bright stars give the cluster its alternative name *Seven Sisters*, a name derived from Greek mythology. There are also much larger associations of stars, known as *globular clusters*, containing many hundreds of thousands of stars. One example, with the rather unromantic name M13, is shown in Figure 1.4; individual stars cannot easily be seen in the heart of the cluster but are visible in the outer regions.

Figure 1.3 The Pleiades, an open cluster.

Figure 1.4   The globular cluster M13.

Actually, there is a sense in which the Sun can be considered as a member of a cluster, the cluster being the *Milky Way galaxy* which is about 100,000 light years across from one side to the other. This is a collection of one hundred thousand million stars forming a recognisable association that is well separated from anything else. It contains field stars like the Sun, isolated binary stars, many open clusters, many globular clusters, clouds of gas and dust and many exotic objects such as neutron stars and black holes — of which more later. The space between these objects is known as the *interstellar medium* (ISM) — a crude description of which is that it is nearly nothing — but not quite. In a volume of the ISM the size of a sugar cube there will typically be one hydrogen atom. In the same volume of the air that you breathe there are about 10^{20} nitrogen or oxygen atoms. A very important component of the ISM is dust. This dust is in the form of particles less than one micron (one millionth of a metre) in diameter. A layer of 50,000 of them would comfortably fit in the dot above the letter *i*. If we consider a cube of the ISM of side one kilometre then that volume would contain just one dust particle! So little — is it even worth mentioning? Yes it is, because this dust plays a vital role in many astronomical processes. In particular it is the stuff that we human beings are made of!

Figure 1.5   The galaxy NGC 6744 — very much like our own Milky Way galaxy.

In saying that the Milky Way is well separated from everything else, we were implying that there are other things from which it is separated. These other things are other galaxies — some like the Milky Way, some bigger, some smaller, some of similar shape and some very different. The structures of nearer galaxies can be clearly seen with large telescopes, and one that resembles the Milky Way, the spiral galaxy NGC 6744, is shown in Figure 1.5. At their greatest distances, hundreds of millions of light years away, galaxies are seen as faint, fuzzy objects. The more powerful the telescope we use, the more galaxies we can see at ever greater distances. These bodies constitute for us the 'observable Universe', estimated to contain 10^{11} galaxies.

How did the Universe and all the objects in it — galaxies, black holes, stars, planets, satellites and many other kinds of object — come into existence? In particular, how did it come about that I am here looking through my study window and thinking about these things?

The Universe

Chapter 2

Christian Doppler and His Effect

2.1 Waves, Frequency and Wavelength

A piano keyboard is an arrangement of white and black keys strung out in a line. A key struck towards the left-hand edge of the keyboard gives a low, booming note — described as being of low pitch. At the right-hand edge the note is lighter, more buoyant in tone — described as being of a high pitch. Sound is a wave motion and, when we hear a sound, alternating high and low pressure air disturbances set the eardrum into vibration. These vibrations are processed and converted into electrical signals that are fed into the auditory cortex of the brain.

The term *pitch* is a qualitative way of describing the frequency of a sound wave — the number of vibrations per second. The scientific unit for frequency is the hertz (Hz), or one vibration per second. Low C on the piano is about 33 Hz (vibrations per second), middle C about 261 Hz and high C is 4,186 Hz. The human frequency range for hearing sound is age-related, but is approximately 16 to 20,000 Hz. Dogs can hear higher frequencies, so a dog whistle that makes no sound to the human ear, is effective in communicating with dogs. The large musical organ in Sydney Town Hall in Australia has huge pipes. The longest (about 20 metres in length) emits vibrations at 8.4 Hz, which are heard not so much as a musical note but rather as time-resolved periodic thumps.

The wavelength of a sound wave is the distance between the high pressure regions (or the low pressure regions) as the sound wave travels through the air (Figure 2.1).

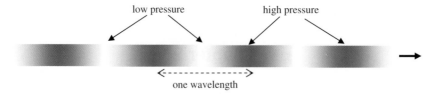

Figure 2.1   The propagated high and low pressure regions that constitute a sound wave. A wavelength is the distance between successive high (or low) pressure regions.

The product of the wavelength and frequency is the speed of propagation of the sound wave, i.e.

$$speed = frequency \times wavelength.$$

The speed of sound in air is about 330 metres per second, so that if the frequency is 110 Hz the wavelength of the sound is 3 metres ($110 \times 3 = 330$). The large pipe in the Sydney Town Hall organ has a length of one half of a wavelength so that the sound it emits has a wavelength of approximately 40 metres. Consequently the frequency it emits (speed ÷ wavelength) is about $330/40 = 8.25$ Hz, close to the 8.4 Hz previously quoted.

2.2 The Doppler Effect and Sound Waves

Emergency vehicles — ambulances, fire-engines and police cars — use sirens that emit a loud and high-frequency undulating sound that alerts other drivers, and pedestrians, of their presence so that a path is cleared to facilitate their rapid progress. A well-known phenomenon is the change of pitch of such sirens between when the vehicles are approaching the listener and when they are receding. As the vehicle approaches so the siren has a higher pitch than that it would have if it were at rest with respect to the listener, and when departing it has a lower pitch. The effect can be reproduced by saying 'ee–er' — try it and see. This phenomenon was first explained scientifically by the Austrian mathematician Christian Doppler (Figure 2.2) and is known as the *Doppler Effect*.

Figure 2.2 Christian Doppler (1803–1853).

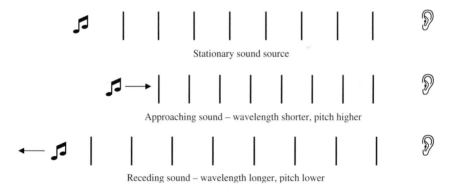

Stationary sound source

Approaching sound – wavelength shorter, pitch higher

Receding sound – wavelength longer, pitch lower

Figure 2.3 The Doppler Effect.

When the source of sound approaches, to the hearer the waves effectively become compressed, the wavelength becomes less and hence the heard frequency (pitch) is higher. Conversely, when the source recedes the waves are effectively stretched out, the wavelength becomes longer and the heard frequency is lower. A schematic representation of this behaviour is illustrated in Figure 2.3.

Theory shows that the fractional change of wavelength equals the ratio of the speed of approach, or departure, to the speed of sound. For example, if an ambulance approaches at 33 metres per second (approximately 120 kilometres per hour or 75 miles per hour) then its approach speed is one tenth of the speed of sound. Hence the wavelength is shortened by one tenth, so if the original wavelength was 1 metre (frequency 330 Hz) it would become 0.9 metres (frequency 367 Hz). When the ambulance is receding then the wavelength is lengthened by one tenth, that is it changes to 1.1 metres, corresponding to a frequency of 330/1.1 = 300 Hz. As the ambulance passes by so the frequency changes from 367 to 300 Hz — a very noticeable effect.

2.3 The Doppler Effect and Astronomy

Just as the Doppler Effect applies to sound so it applies to light, another kind of wave motion. Light is an electromagnetic radiation, which involves the coordinated vibrations of electric and magnetic fields. Electromagnetic radiation covers an enormous range of wavelengths, forming a continuum although we conventionally give different names to different sections of the electromagnetic spectrum (Figure 2.4). At the shortest wavelengths we have γ-rays which are really very hard and penetrating X-rays. Just beyond the blue end of the visible spectrum there is the ultraviolet (UV) region and just beyond the red end of the visible spectrum there is the infrared (IR) radiation. The long wavelength end of the infrared region is sometimes called heat radiation. At longer wavelengths than infrared we enter the radio region which includes the range of wavelengths and frequencies used for radio and television transmission. Radiation over the whole of this range of wavelengths is emitted by one type or other of astronomical object and astronomers have invented specialised instruments that can detect them — from γ-ray detectors to radio telescopes — and, in many cases, even produce images, just as the eye produces an image from visible light. It is interesting to note in Figure 2.4 what a tiny fraction of the electromagnetic spectrum corresponds to the visible light that enables us to see.

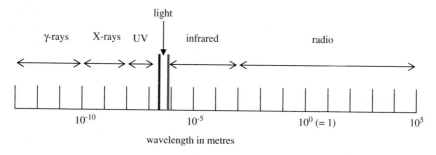

Figure 2.4 The electromagnetic spectrum.

Figure 2.5 A solar spectrum.

All electromagnetic radiation travels with the speed of light, 300,000 kilometres per second, so we are not going to detect Doppler shifts of light wavelengths in normal everyday life. A police car, even in full chase, would be travelling at one ten-millionth of the speed of light so we would not expect to see any colour changes in the blue lamp on its roof. However, astronomical objects move at higher relative speeds — for example, some of the stars we see can be moving towards, or away from, us at speeds of around 30 kilometres per second, just one ten-thousandth of the speed of light. Now, changing the wavelength of light by one ten-thousandth of its original value is not going to be detectable to the eye, but nature gives us a hand in measuring such small differences.

If light from the Sun is spread out with a prism, we see a spectrum, like a rainbow, going from violet to red. However, when we look more carefully we see that there are various dark lines in the spectrum corresponding to some specific wavelengths (Figure 2.5).

These are *Fraunhofer lines* and are due to the presence of different chemical elements in the outer regions of the Sun. The light from the Sun is generated mainly in a thin layer called the *photosphere*. There is insufficient material above the photosphere to generate much light and light generated below the photosphere gets absorbed before it can escape. However, there *is* enough material above the photosphere to give significant absorption. Thus hydrogen, because of its electronic structure (Figure 6.8), preferentially absorbs a number of discrete wavelengths and the absence of these wavelengths in the spectrum tells us that hydrogen is present. That is no surprise — some 80 per cent of the Sun is made of hydrogen. The first experiments showing Fraunhofer lines were carried out by Joseph Norman Lockyer (1836–1920) who, in 1868, attached a spectrometer to a telescope, which spread out the light into a spectrum. Nearly all the Fraunhofer lines could be associated with known elements, the spectral lines of which could be reproduced in laboratory experiments. However, there was no match to one very strong line in the yellow region of the solar spectrum and, in 1870, Lockyer suggested that this was due to an as-yet unknown element, which he named helium (the Greek for the Sun is Ηλιος – *helios*). Twenty-five years later, in 1895, this prediction was confirmed when William Ramsey (1852–1916) extracted helium that was trapped inside the mineral cleveite. Helium is the second most common element in the Sun and in the Universe at large; it is amazing that this very common element was discovered on an astronomical body, the Sun, before it was found on Earth!

By the optical technique *interferometry*, the wavelengths of spectral lines can be measured very accurately. We know that the laws of physics and chemistry are the same on stars as they are on Earth so if we measure the wavelength of a spectral line of, say, iron in the laboratory then we know that it is the wavelength emitted by iron on the star. If the star is moving away from the Earth then the Doppler Effect will give a measured wavelength which is longer — towards the red end of the spectrum — and we say that the light has been *red-shifted*. Similarly, if the star moves towards the Earth, so that the measured wavelength is shorter than the laboratory measurement, then we say that the light is *blue-shifted*. These tiny shifts of wavelength can be measured very accurately by interferometers and the best instruments can give estimates of speed with an accuracy approaching one metre per second.

The measurement of optical Doppler shifts to measure speeds towards or away from the Earth is a very powerful astronomical tool and is used by astronomers in many different contexts. In the next and subsequent chapter we shall see how its use has led to a revolution in our understanding of the Universe.

Chapter 3

Measuring Distances in the Universe

3.1 The Parallax Method

Try the following experiment. Position yourself far from some distant panorama, such as a bank of trees. Hold a finger vertically, at arm's length and symmetrically between your two eyes. Close one eye and then the other and observe how your finger moves relative to the distant scene. This phenomenon, known as *parallax*, is the basis of measuring the distances of nearer stars.

In the astronomical application of parallax, the equivalence of the finger is the nearby star and that of the distant scene is the background star field consisting of stars which are very distant. To see the near star move against the background stars we now need two different viewing positions — equivalent to first closing one eye and then the other. These viewing positions are provided by the Earth's orbit around the Sun; if the nearby star is observed at times six months apart then the distance between the viewing points is 2 astronomical units, or about 300 million kilometres. This is illustrated in Figure 3.1, which is not to scale because it is not possible to show the true relative positions of near and far stars, the distances of which could be in the ratio 1:10,000 or even greater. For the maximum parallax effect, the observations are taken at points such that the line AB is perpendicular to the direction of S.

The points A' and B' are noted in relation to the background stars. Since the distance AB is known then, if the angle α can be determined, the distance of the star, S, can be found. When the distant stars are very much further away than S the angle between A' and B' as seen from

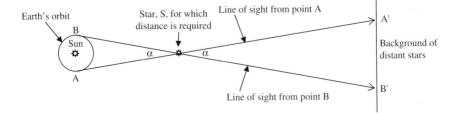

Figure 3.1 The basis of the parallax method.

either A or B, which is easily measured, will be virtually the same as α and so the distance of S can be found. There is an apparent problem in that stars move, so that in the six months between taking the observations at A and B the star S has shifted its position. However, this is not really a problem and is solved by taking three observations — at A, at B and then at A again. Taking these three measurements has the advantage that not only can the distance of S be found but also the component of its speed in a direction perpendicular to the line of sight.

 Using the parallax method with ordinary telescopes, the distances of a few thousand stars out to about 100 light years can be found. With terrestrial telescopes that use *adaptive optics*[a] to remove the shimmering effect when light moves through the atmosphere, this range is increased to 200 light years. The Hipparchos satellite, launched by the European Space Agency in 1989 specifically to make parallax measurements, has extended the range to about 600 light years, thus enabling the distances of approximately one million stars to be found. While this may seem a large number it is only one in one hundred thousand of the stars in the Milky Way galaxy. To extend distance measurements to all the stars in the galaxy, some other method must be used.

3.2 Main-Sequence Stars

All stars have a life cycle with a birth stage, a mature stage and a death stage, all of which will be discussed later. For Sun-like stars the mature

---

[a] A full description of adaptive optics is given in M.M. Woolfson (2011), *The Fundamentals of Imaging: From Particles to Galaxies* (London: Imperial College Press).

stage is known as the *main sequence*, during which the star is producing energy by converting hydrogen into helium by nuclear processes. The Sun is about 4,600 million years into its main sequence and will remain in it for another 5,000 million years, after which it will undergo the processes that we can think of as the death of a star. During the main sequence stage the state of the star, in terms of its brightness and other physical characteristics, remains roughly constant. An important consideration is that, relative to the other stages of its existence as a bright object, the main sequence is long-lasting so that a high proportion of the stars we observe are main-sequence stars.

A characteristic of a star that we can always estimate, regardless of its distance, is its temperature. If an iron object is heated to two or three hundred degrees centigrade it does not emit any visible light. It does emit electromagnetic radiation but this is heat radiation that we can feel but cannot see. At a higher temperature the iron begins to glow a dull red colour. Increasing the temperature further changes the colour to orange, then yellow, then white and finally white with a bluish tinge. The distribution of wavelengths radiated from a hot body, which gives the overall colour effect, is dependent on its temperature. A star moved ten times further away will decrease in brightness, but its colour will not change. Actually, astronomers estimate the temperature of a main-sequence star not by looking at the wavelengths it emits but rather by looking at the Fraunhofer absorption lines seen in Figure 2.5. The distinctness of individual lines, coming from different types of atom in the star, varies with temperature. The lines from different chemical elements change in different ways — for example, as temperature increases so a particular hydrogen absorption line might become weaker while another, from iron, becomes stronger. By comparing the relative strengths of many lines (Figure 3.2), astronomers can assess the temperatures of main-sequence stars, with an accuracy of a few degrees in favourable cases.

When driving at night the headlights of oncoming vehicles can be very dazzling. However, this is only a problem when the vehicle is close; when seen at a distance the same headlights are much less troublesome. If all headlights were exactly the same in their

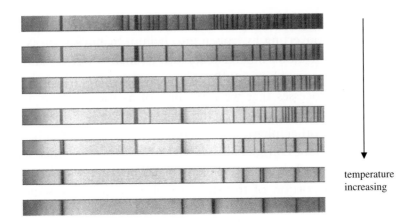

temperature
increasing

Figure 3.2   The change in the pattern of spectral lines with temperature. The lowest
temperature (at the top) is about 3,000°C and the highest temperature
(at the bottom) about 30,000°C.

intrinsic brightness (the rate at which they emit light energy) then,
from their *apparent* brightness when we observe them, it would be
possible to estimate their distances. The apparent brightness falls by
the square of the distance — if the distance is doubled then the
apparent brightness goes down by a factor of four and if the distance is trebled then the apparent brightness goes down by a factor
of nine.

Using this principle, the study of stars within parallax range shows
that all main-sequence stars with the same temperature have closely
the same intrinsic brightness. This property enables the distances of
main-sequence stars to be found outside the parallax range. The pattern of absorption lines enables the fact that it *is* a main-sequence star
to be established and also indicates its temperature and hence its
intrinsic brightness. Now an instrument is used that measures the
energy received from the star, which gives the apparent brightness.
Finally we find the distance corresponding to that *apparent* brightness, given the known *intrinsic* brightness. This method can be used
for stars well outside the parallax range, out to about 20,000 light
years, although with decreasing accuracy with greater distance as the
absorption lines become increasingly indistinct. If the object under

observation is a galactic cluster containing many main-sequence stars then, by using the aggregate information from the whole cluster, distance estimates can be made better than that from a single star.

3.3 Using Cepheid Variables

The groundwork for the next step in measuring even greater distances was due to observations first carried out in the 18th Century on variable stars — stars whose brightness varies in a periodic way. By the middle of the 18th Century there were a few stars that were known to vary in brightness but nobody had measured the variation — it was just a matter of note. The pioneer in the systematic study of such stars was the English astronomer John Goodricke (Figure 3.3), who, despite the social handicap in those days of being a deaf mute, became educated, took up astronomy and discovered a number of important variable stars. The first of these, observed in 1782, was Algol, the fluctuating light curve of which was explained by Goodricke. Algol is an eclipsing binary system, consisting of a pair of stars, of similar size, but with one much brighter than the other. The two stars circle each other with the plane of their orbits such that, for an observer on

Figure 3.3   John Goodricke (1764–1786).

Earth, one star can move in front of the other so that the light from the one behind is totally or partially obscured. For most time the light from both stars is seen; when the brighter star is at the front there is a small diminution of brightness but when the dimmer star is in front there is a large and dramatic fall in the intensity.

For finding distances, the most significant discovery by Goodricke was the variable star δ-Cephei, which fluctuates in brightness with a period of about 5.4 days. It is believed that Goodricke's early death, at the age of 21, was due to pneumonia, contracted as a result of exposure when observing this star. This star was the prototype of stars, called *Cepheid variables*, which differ in their average brightness and period. Many of these stars are within the parallax range so their distances are known and thus for any particular Cepheid variable the intrinsic average brightness can be found. In 1908 a Harvard astronomer, Henrietta Leavitt (Figure 3.4), found a relationship between the average brightness of Cepheid variables and their periods (Figure 3.5). The figure shows that Cepheid variables can be very bright stars, up to more than 30,000 times brighter than the Sun, and so they can be

Figure 3.4　Henrietta Leavitt (1868–1921).

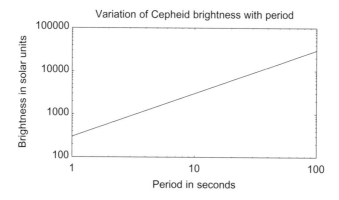

Figure 3.5 The relationship between the brightness of a Cepheid variable star and its period.

seen at great distances. In particular they can be clearly seen in the outer regions of some nearer galaxies. If the period of a Cepheid variable in a galaxy is measured then, from Figure 3.5, its intrinsic average brightness may be inferred. Hence, from the measured apparent brightness the distance of the galaxy can be estimated. In this way distances can be measured out to about 80 million light years, a large distance but still far from the boundary of the observable Universe.

3.4 Spinning Galaxies

We have seen that the secret of measuring distances is to have some object of determinable intrinsic brightness that can be seen from Earth. The further the object the brighter it must be to be seen, so if we are to extend well beyond the Cepheid variable range we need a much brighter object. The next objects we can use are spiral galaxies, similar to our own Milky Way. From the way that stars move within the Milky Way we know that our galaxy is spinning, making a complete revolution in 200 million years. While that is a slow rotation, because of the size of the galaxy, stars on opposite sides of the galaxy have a relative speed of about 1,000 kilometres per second. Someone looking at our galaxy edge-on would see the stars on the right-hand side of the galaxy moving away while those on the left-hand side were

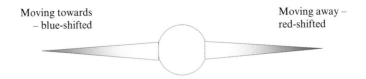

Moving towards
– blue-shifted

Moving away –
red-shifted

Figure 3.6   The spectral line shifts from a rotating spiral galaxy seen edge-on.

moving towards him (Figure 3.6) — assuming for now that the centre of the galaxy is at rest with respect to the observer. This would give a red shift to the stars on the right (moving away from the observer) and a blue shift to stars on the left (moving towards the observer). The observer would not be looking at the light from individual stars but at the aggregate light from all stars from various parts of the galaxy. The light from the centre of the galaxy would have no shift and as one looked further to the right the red shift would steadily increase and there would be a similar increase in blue shift as one looked further and further to the left. Looking at one spectral line, of the kind shown in Figure 2.5, it would be spread out due to the different wavelength shifts from different parts of the galaxy. The width of this spread would give a measure of the rotational speed of the galaxy and would not be altered by overall motion of the galaxy towards or away from the observer. The centre of the line would be red-shifted or blue-shifted but the width of the line would be unaffected.

Analysis of the spread of spectral lines for spiral galaxies within the range of Cepheid variables shows that the width of the spectral lines (dependent on the rotation speed) is closely related to the brightness of the galaxy. The physical link between brightness and rotation speed is the galactic mass. The mechanics of a spiral galaxy gives rotation speed increasing with total mass. However, the greater the galactic mass the greater the number of stars it contains and so the brighter it will be. Through this linkage we see that faster rotation is correlated with greater brightness; the connection between these two quantities is well established and is known as the *Tully–Fisher relationship*. By measuring the spread of the spectral lines of a distant spiral galaxy its intrinsic brightness can be estimated, and its apparent brightness then gives its distance. Using this tool, distances can be measured out to about 600 million light years, beyond which distance the images of galaxies are too faint to give a good estimate of the spectral-line widths. Now we have stretched our distance measurements a long way — but there is still far to go before we reach the limit of the observable Universe.

3.5 Using Supernovae as Standard Sources

We have already mentioned that the Sun is halfway through its main-sequence stage. At the end of its main-sequence existence it will shed layers of material, forming a planetary nebula (Figure 3.7) and eventually settle down as a *white dwarf*. These are strange objects, of Earth-size but with about the mass of the Sun. A teaspoonful of white-dwarf material would have a mass of about 10 tonnes! A star much heavier than the Sun will undergo a more violent end. It will complete its main-sequence existence in a violent explosion known as a *supernova*, when for a period of some weeks it can be as bright as ten billion Suns. There are other ways in which violent supernovae outbursts can occur so there are several kinds of supernovae, which

Figure 3.7 A planetary nebula. This is a complete shell of material although, in projection, it looks like a ring.

can be individually recognised by what they show in their spectra. An important type of supernova occurs when a white dwarf is in a binary system with another kind of star called a *red giant*. This is a comparatively cool, but very large, star in a stage of development between being in the main sequence and becoming a white dwarf. When the Sun goes through the red-giant stage of its existence it will expand to just about encompass the Earth. Red giants tend to shed material and if there is a white dwarf in a binary relationship with it then the shed material, or some of it, will attach itself to the white dwarf. A white dwarf, if left alone in isolation, will quietly cool down until it ceases to shine at which stage it becomes a *black dwarf*. However, if while in the white dwarf stage it steadily gains mass then, when it reaches a critical mass known as the *Chandrasekhar limit*, about 1.44 times the mass of the Sun, it becomes unstable and explodes to give what is known as a type 1a supernova. Because of the way they come about — they all reach the same critical mass — all type 1a supernovae are very similar and in particular the peak brightness is the same from one type 1a supernova to another.

Type 1a supernovae are very bright and can be seen in distant galaxies. Because some type 1a supernovae have occurred within galaxies for which Cepheid variable distances are available their maximum intrinsic brightness is known. This means that when such a supernova is seen in a galaxy outside the Cepheid variable limit, the distance of that galaxy can be determined. In this way distances can be determined out to 3,000 million light years but, even so, we have still not yet reached the limits of the observable Universe.

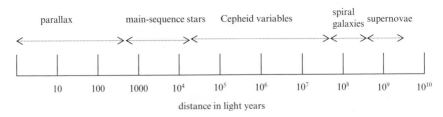

Figure 3.8 Techniques for measuring distances in the Universe. Methods can be used below the limits shown, but not above.

The limits of the various techniques for measuring distances are illustrated in Figure 3.8. Once through the parallax region, the boundaries of measurement are moved forward by finding new sources, the greater intrinsic brightness of which can be found by comparison with the sources from the closer-in region. Supernovae are the ultimate source for extending the boundaries in this way. Something else is needed in order to reach the edge of the observable Universe.

Chapter 4

Edwin Hubble's Expanding Universe

4.1 Galaxies, Clusters and Superclusters

By using information about astronomical bodies within one distance to develop a way of extending distance estimates further, we can determine the distances of galaxies up to 3,000 million light years away. It is a strange thought that, when we see a galaxy at the limit of this range, we see it as it was 3,000 million years ago, when only primitive life forms inhabited the Earth. If intelligent life exists within that distant galaxy that can observe the Milky Way, then they see our galaxy as it was 3,000 million years ago. Given such an advanced technology that they could see details of life on Earth then they could beam information towards Earth so that our descendants, some 3,000 million years in the future, would then receive a first-hand account of what life was like, 3,000 million years ago to us and 6,000 million years ago to them. However, I am musing again — I must keep on track!

When two close galaxies are seen in a telescope's field of view, we cannot tell *ab initio* whether they are actually close together or well separated along the line of sight. The apparent sizes of the galaxies might give some clue as to their relative distances, but not reliably since galaxies are not all similar in size. The direction *and distance* of a galaxy determine its exact position in space and the positions of galaxies in three dimensions reveal that there are *clusters of galaxies*. A number of galaxies, mostly well separated from each other, form a cluster, separated from other clusters of galaxies by distances much larger than the separations of galaxies within each cluster. Seen

31

Figure 4.1   A stereoscopic pair of images that shows a representation of two clusters of galaxies separated along the line of sight. By focusing your eyes to a point beyond the page you should be able to merge the images to get a three-dimensional view of the arrangement. If you have difficulty in merging the images then begin with the figure a few centimetres from your face and then slowly move it away until the figure comes into focus. It is important to keep the figure horizontal.

through a telescope, galaxies seem to be situated in random directions; it is only when the distances are found that the clustering becomes apparent. The left-hand side of Figure 4.1 shows a notional part of the sky containing 20 galaxies. There is no obvious relationship between them. Now, on the right-hand side of the figure we introduce the element of distance by reproducing a stereoscopic partner to the left-hand side. With a little practise you should be able to look at the complete figure and to merge the two sides so that you see a three-dimensional image, which shows that the galaxies are in two groups, each consisting of ten galaxies, separated in distance along the line of sight. Of course we cannot see real galaxies in stereoscopic view, which is similar to the parallax method, since it involves seeing the same scene from two locations; in practice the distance information comes from the methods described in Chapter 3.

The Milky Way is part of a local cluster of more than 40 galaxies known as the *Local Group*, with diameter six million light years, 60 times the diameter of the Milky Way. Most galaxies in this group are considerably smaller than the two largest galaxies — the Milky Way and *Andromeda*, another spiral galaxy. A neighbouring galaxy, the *Large Magellanic Cloud*, situated a mere 160,000 light years from Earth, hit the astronomical headlines in 1987 when a supernova occurred within it. Although supernovae are not particularly rare events — they occur in distant galaxies on a regular basis — having one close at hand *is* quite rare and this particular one was subjected to close scrutiny.

A cluster of galaxies, like a cluster of stars or the contents of a single galaxy, is bound together by the mutual gravitational attractions of its constituent bodies. The individual galaxies in a group move relative to each other, slowly changing their positions, speeds and directions of motion. By Doppler-shift measurements, the velocities along the line of sight of the other galaxies within the Local Group can be found, and it is found that some are moving towards the Milky Way while others are moving away from it. When the velocity along the line of sight is towards the Milky Way, that does not mean that a collision between the galaxies is going to occur. It just means that the distance between the Milky Way and the other galaxy is decreasing but the closest approach distance could be quite large (Figure 4.2).

Clusters of galaxies are not the largest scale of organisation of the Universe. Clusters of clusters, known as *superclusters,* also exist. The superclusters, again bound together by gravity, are separated from other superclusters by distances that are large compared with the distances between clusters within each individual supercluster. A typical supercluster is about one hundred million light years across; in Figure 4.3 there is a schematic arrangement in two dimensions, not to scale, of galaxies, clusters of galaxies and superclusters. The Local Group, the cluster in which the Milky Way is situated, is a member of the *Virgo Supercluster.* The scales of these structures strain human

Milky Way

Figure 4.2 A galaxy of the Local Group getting closer to the Milky Way without being on a collision path.

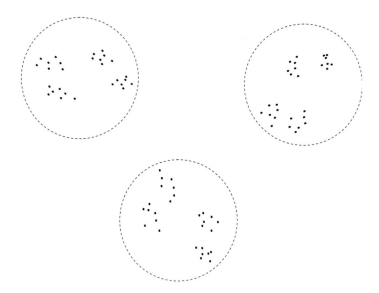

Figure 4.3   A schematic two-dimensional representation of galaxies (individual symbols), clusters of galaxies and superclusters of clusters (ringed).

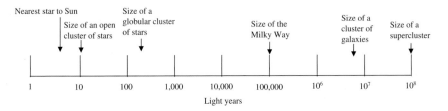

Figure 4.4   The hierarchy from the distance of the closest star to the Sun to the size of a supercluster.

imagination; in Figure 4.4 we give a representation of the scaling hierarchy of distances from the separation of stars to the size of a supercluster.

4.2 Hubble's Law

A prominent worker in the field of measuring galactic distances was the American astronomer Edwin Hubble (Figure 4.5). Hubble not only found their distances but also used the Doppler-shift method to

Figure 4.5   Edwin Hubble (1889–1953).

determine their velocities along the line of sight. Some of the velocities he found were very large — a considerable fraction of the speed
of light. For such velocities Doppler's original equation does not
apply and, instead, one must use a different equation — the *relativistic Doppler-shift equation*. Figure 4.6 shows this equation in graphical
form, where the ratio of the observed wavelength to the wavelength
of the emitted radiation (vertical scale) is shown as a function of the
velocity expressed as a fraction of the speed of light (horizontal scale).
A positive velocity represents motion away from the observer and a
negative velocity motion towards the observer. For velocities that are
small compared with the speed of light the result is almost that originally given by Doppler, but the result is very different when the velocity is comparable to the speed of light. An application of the original
Doppler equation would indicate that, for a body moving away from
an observer with the speed of light, the observed wavelength would
be twice that emitted. The relativistic equation indicates that, in fact,

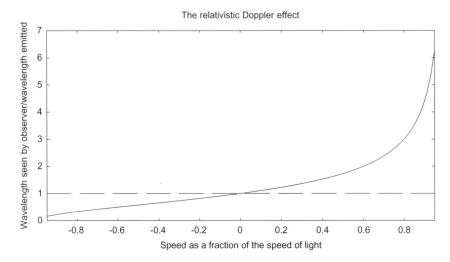

Figure 4.6   The relativistic Doppler effect for speeds comparable to the speed of
light.

the observed wavelength would be infinite. These large wavelength
changes caused a number of problems in early analyses of stellar spectra because spectral lines that were normally invisible could be seen.
For example, an ultraviolet spectral line of wavelength 200 microns[a]
could be shifted to 500 microns and so it would be seen as a line in
the green part of the spectrum. The early workers were not expecting
relativistic velocities — all they knew at the time was that lines were
appearing in the spectrum that they could not account for. Only when
they realised that they were dealing with very high velocities, and
hence large wavelength shifts, could they make any sense of their
observations.

Hubble's results were spectacular and led to the area of astronomy we call *cosmology*, the study of the Universe. He found that,
beyond the Local Group, the galaxies he observed were *moving away
from the Earth with speeds that were proportional to their distance*. This
means that if a galaxy at a distance of 300 million light years was
receding at a speed of 6,000 kilometres per second then a galaxy twice

---

[a] 1 micron is one millionth of a metre.

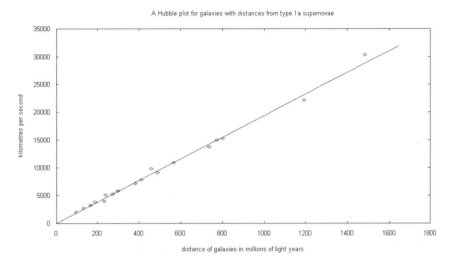

Figure 4.7 An illustration of Hubble's law.

as far away, at a distance of 600 million light years, would be receding at 12,000 kilometres per second. The observation that speed of recession is proportional to distance is known as *Hubble's law*. Figure 4.7 shows the distances of galaxies, found from observations of type 1a supernovae, against the speed of recession and it will be seen that the fit of points to a straight line is quite good.

If Hubble's law is valid at any distance, then the distances of very distant galaxies, in which type 1a supernovae would not be observable, can be found by determining their speeds of recession by measuring Doppler shifts. Again, if this rule applies to all galaxies, regardless of their distance, then the inescapable conclusion is that *the Universe is expanding.* This simple statement has profound philosophical implications, some of which we shall be considering later.

We conclude our description of the expanding Universe by considering an important question raised by Heinrich Olbers (1758–1840) in 1823, called *Olbers' paradox.* The question is: Why is the sky dark at night? It seems a silly question; we just accept that, when the Sun sets, the sky is dark. However, the sky is not completely dark; it contains stars, some seen directly with our unaided eyes and many more stars and distant galaxies seen with the assistance of a telescope.

If we lived in an infinite Universe, uniformly filled with stars and galaxies, then it can be shown theoretically that a star would be seen along *any* line of sight. That means that the sky should be seen full of the overlapping images of stars and hence that the whole sky should be as bright as a star — which is clearly not true.

The answer to this paradox comes from the expansion of the Universe. Light that leaves a distant receding galaxy with a particular wavelength is red-shifted and so arrives at Earth with a greater wavelength. The greater the wavelength of the light, the less energy it has, and so distant galaxies are less bright than those near at hand. Another effect is that as the galaxy recedes so the space between the galaxy and the Milky Way increases and more emitted light exists as radiation in that space. Hence, even without the Doppler effect, the rate of arrival of light is less than the rate of emission, again leading to the distant galaxy appearing less bright. These factors explain Olbers' paradox. Without the expansion of the Universe and the Doppler effect, the whole sky would be radiating like the surface of the Sun, and so make our planet uninhabitable.

Chapter 5

A Weird and Wonderful Universe

5.1 The Classical Universe

Astronomical measurements reveal that the Universe is immense, is expanding and contains objects whose relative speeds are large fractions of the speed of light. Because of the immense distances between objects in the Universe and the finite speed of light, we never know the state of a body when we observe it, so time is another factor that must be included in attempting to understand the structure of the Universe.

The instinctive feeling we have about space and time was formulated by Isaac Newton (Figure 5.1). In his famous publication *Principia*[a] he stated:

- Absolute, true and mathematical time, from its own nature, passes equably without relation to anything external.
- Absolute, true and mathematical space remains similar and immovable without relation to anything external.

In this view of space and time, the place and time of any event will be seen by all observers to be the same. This is intuitively acceptable; if one person said that an event had happened in Trafalgar Square at 2.00

---

[a] *Principia* is a shortened form of the full title *Philosophiae Naturalis Principia Mathematica* (Mathematical Principles of Natural Philosophy) in which Newton gives his laws of motion, of universal gravitation and of planetary motions.

Figure 5.1   Isaac Newton (1642–1727).

pm on 1st January 2007 while another claimed that the same event had happened elsewhere at another time then we would find that confusing. We believe those things that conform with our experience. A child playing ball knows nothing of Newtonian mechanics but does know how to toss a ball upwards so that it can be caught by a friend; he accepts what experience tells him is true. Similarly we know that light does not bend round corners and that a flame is hot and should not be touched. There is survival value in accepting the fruits of our experience.

When dealing with tiny entities such as atoms or electrons, or huge entities like the Universe, or bodies moving at appreciable fractions of the speed of light such as distant galaxies, then experience cannot guide us and we should expect to find behaviour patterns outside our experience that we do not properly understand. In the world we understand the distance between London and Edinburgh is 640 kilometres and is the same to everyone else. When the small hand of a clock moves from one to two then an hour has passed and it cannot be otherwise to anyone else. That is what Newton said.

At the end of the 19th Century, physics rested on two foundations that dealt with different aspects of the physical world. The first of

Figure 5.2   James Clerk Maxwell (1831–1879).

these was Newtonian mechanics, which described the way that objects moved and included the action of gravity that exerted force at a distance without the need for intervening physical material. The second, less well known to the layman but of equal importance, was the work of the Scottish scientist James Clerk Maxwell (Figure 5.2). Despite his short life Maxwell made significant contributions in many areas of physics, the most important of which explained the behaviour of, and interactions between, electricity and magnetism and showed that light was an electromagnetic wave. These two foundations, by Newton and Maxwell, seemed to be independent of each other, but together dealt with all aspects of physics.

5.2 The Relativistic Universe

Experiments carried out at the beginning of the 20th Century seemed to be in conflict with the classical world described by Newton and Maxwell, in which particles and electromagnetic waves had separate existences. Electrons, identified as negatively-charged particles, could, in some circumstances, behave like light — electromagnetic radiation, identified as a wave motion, with wave-like behaviour. Conversely, light could sometimes behave like particles, in bullet-like entities called *photons*. How electrons and light behaved depended on the experiment being performed. If an experiment was designed to detect light as photons then the light very obligingly behaved in that way. Similarly, an electron microscope uses electrons to form an image just as an ordinary light microscope uses light; the electron microscope is designed for electrons to show their wave-like properties and they conveniently do so.

Other experiments early in the 20th Century showed that light behaved differently from other kinds of wave or moving object. A car moving on a motorway at 110 kilometres per hour relative to an observer on a bridge would be moving at a speed of 10 kilometres per hour relative to an observer in another car travelling in the same direction at 100 kilometres per hour; speed can be different relative to different observers. A water wave moving at 15 kilometres per hour along a canal relative to an observer sitting on the bank would appear to be stationary to someone on a bicycle travelling at 15 kilometres per hour in the same direction. However, various experiments with light gave the same speed of light for all observers, which was a source of some puzzlement and controversy. The situation was resolved by Albert Einstein[b] (Figure 5.3), who started with the proposition that the speed of light was the same to all observers and then found out what the consequences were of that proposition. This led to Einstein's 1905 *Theory of Special Relativity*, which gave conclusions in conflict with the Newtonian world that forms part of our experience.

Einstein's world was a strange one. A person in a train travelling at one half of the speed of light from London to Edinburgh would assess the distance he travelled as 554 kilometres. Someone standing outside and observing a clock within the train would notice that it was

---

[b] Einstein was awarded the Nobel Prize in Physics in 1921, but not for the work for which he is best remembered, that on relativity, which aroused so much controversy at the time that it was thought 'dangerous' to recognise it in so outstanding a way. The prize was awarded for his work on the photoelectric effect, which showed that in some circumstances light can behave like particles.

Figure 5.3   Albert Einstein (1879–1955).

running rather slowly, seeming to take 69 seconds for the hands to move forward one minute. The person by the trackside would find the distance from London to Edinburgh as 640 kilometres while the person on the train would conclude that the train's clock was keeping perfect time. Another thing to trouble our trackside observer is that the train would seem to be strangely compressed. London to Edinburgh trains were known to be 150 metres in length but the length would be only 130 metres as estimated from a photograph taken as it passed by.

An even stranger conclusion from Einstein's theory is the *twin paradox*. One of a pair of twins makes an epic journey by spacecraft to and from a nearby star at three-quarters the speed of light. He returns after experiencing 30 years according to the clock on the spacecraft. When he returns he is told that his journey took 45 years according to Earth clocks and, indeed, the Earth-bound twin is 15 years older than his much-travelled brother (Figure 5.4). You may think that this conclusion is crazy and impossible, but experiments have been done that illustrate the principle of the twin paradox. Elementary charged particles, called mesons, decay into something else at a certain rate. When these particles are accelerated to a very high speed — very close

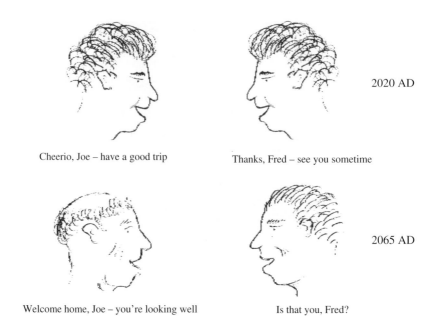

2020 AD

Cheerio, Joe – have a good trip

Thanks, Fred – see you sometime

2065 AD

Welcome home, Joe – you're looking well

Is that you, Fred?

Figure 5.4 The twin paradox. The traveller comes home younger than his twin brother.

to the speed of light — the mesons decay much more slowly. Relating ageing humans to decaying particles, we can say that the moving particles are younger than their twins at rest in the laboratory. Another conclusion from special relativity is the equivalence between energy and mass and that in some circumstances mass can be turned into energy (and *vice versa*!!). Nuclear power generation and nuclear weapons are a manifestation of that conclusion.

Special relativity theory applies to bodies that are not experiencing forces and hence not being accelerated. To deal with accelerating bodies, and the influences that cause acceleration such as gravity, Einstein proposed his *General Theory of Relativity* in 1915. As modelled in this theory, the presence of a massive body distorts a combination of space and time called *spacetime*. Let us see what this means. Figure 5.5(a) shows the path of a particle moving in a straight line defined in a two-dimensional grid. Now, in Figure 5.5(b) the grid is distorted, actually by the use of a mathematical formula but we could

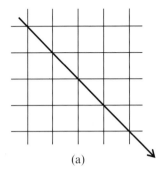
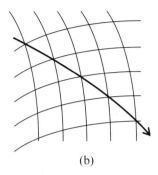

(a) (b)

Figure 5.5 (a) A grid defining position in a two-dimensional space and the path of a particle in the space. (b) A distortion of the grid with a corresponding distortion of the path of the particle.

imagine through the gravitational action of a nearby body, and the path of the particle is also distorted. With a little imagination, we can extend this picture into three dimensions. Positions in space are defined by going up-and-down, from side-to-side and in-and-out of the page. When these lines are distorted, the path of a particle, originally straight, is distorted into a curve. We now go one step further, taking a step that challenges our imagination but is quite straightforward in mathematical terms; that is the power of mathematics — it transcends the limitations of the human mind and imagination. We now consider a four-dimensional space, called spacetime, in which three of the dimensions are the ones in which we live and the fourth is a representation of time. A point in the spacetime grid indicates not only where a particle is situated but also when it was there. Without gravitational or other forces a body will move in a straight line through spacetime at a constant speed. The gravitational effect of a massive body distorts spacetime, and so distorts the path of the body.

We have described, in rather general terms, the theoretical understanding of space and time which existed when Hubble established that the Universe was expanding. Several challenging questions occur — for example, 'If the Universe is the totality of space that exists then what is the Universe expanding into?' Again we are limited by our experience in considering such matters. To establish an idea of what an expanding Universe means, we consider a hypothetical

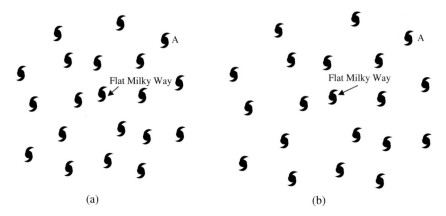

Figure 5.6   (a) Galaxies in the flat universe at time of observation. (b) Positions of galaxies at some later time.

universe that is simpler than the one we inhabit — a two-dimensional universe, inhabited by flatlanders who are completely without thickness and can move only in a plane. A flatland Hubble observed the two-dimensional universe through his two-dimensional telescope and found that it is expanding. Figure 5.6(a) shows a distribution of galaxies in this universe at some time, with the Flat Milky Way at its centre, and Figure 5.6(b) shows the distribution later, when the galaxies have moved further away from the Flat Milky Way with a speed proportional to their distance from it.

Although the expansion has been described as though it was centred on the Flat Milky Way, the view as seen from any other galaxy is that all other galaxies are moving away at a speed proportional to their distance. This is a great relief; it would be incredible if our galaxy was rather special in that it was the only one in which Hubble's law was valid. Another conclusion is that if the flat universe stretched out so far that, with the application of Hubble's law, galaxies were moving away from the Flat Milky Way with more than the speed of light then it would be impossible to see them from Flat Earth. We have already mentioned that light can act like a stream of particles — photons — and the energy of a photon is proportional to the reciprocal of (one over) its equivalent wavelength. A photon emitted from a distant galaxy moving away from the Flat Milky Way at the speed of light would be seen on Flat Earth with an infinitely long wavelength (Figure 4.6). The reciprocal of infinity is zero so the photon would have no energy associated with it and could not be recorded. This would be true for any photon of any initial wavelength leaving the distant galaxy and hence the distant galaxy would be invisible. This raises deep philosophical questions: 'Is there any reality associated with objects that are completely undetectable and cannot affect us in any way' and, 'is the universe defined just by those objects which can be detected?' Let us suppose that galaxy A in Figure 5.6 is just within the bounds of observation from Flat Earth. If galaxy A were at the boundary of the universe, and nothing existed beyond it, then an observer on galaxy A would have a very asymmetric universe to look at, with one half of his sky empty and the other half full of galaxies — but only out as far as the Flat Milky Way — he would not see anything beyond that. If the universe had a circular boundary with the Flat Milky Way at its centre then the Flat Milky Way *would be special* — which is philosophically unacceptable.

The answer to the conundrum, that we want a flat universe in which no galaxy is special, is to introduce a third dimension, which would be as understandable to a flatlander as a fourth space dimension is to us. Instead of the galaxies being arranged in a plane we arrange them on the surface of a sphere (Figure 5.7). It is as though the flat galaxies were arranged on the surface of a spherical balloon and the expansion of the flat universe corresponds to the steady inflation of the balloon. We see

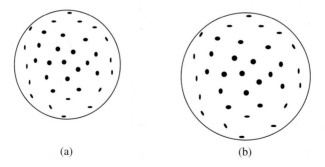

(a)   (b)

Figure 5.7   (a) Flat galaxies arranged on the surface of a sphere. (b) An expanded flat universe.

this model in three dimensions but the flatlanders will only be conscious of a flat universe. Light will follow the curvature of the balloon's surface and, as far as the flatlanders are concerned, the light will be travelling in a straight line (we are ignoring the fact that gravity can bend light, a result from the theory of general relativity). Observers in each galaxy will have a sight horizon limited by those galaxies travelling at a relative speed less than the speed of light and each will therefore consider himself as the centre of the observable universe.

The flatlanders would not have understood our description of their universe, because it involved a dimension outside their comprehension, but it made perfect sense to us. We now describe our own Universe as seen by a being that lives in four-dimensional space. The galaxies occupy a three-dimensional space that has the volume of the surface of a four-dimensional sphere. The three-dimensional space is curved and has no boundaries. All galaxies within the three-dimensional space see themselves as the centre of the Universe. The expansion of the Universe corresponds to inflation of the four-dimensional sphere. If that does not make sense to you then do not worry — the vast majority of people will not really understand it and be able to visualise it in terms of the world, or the Universe, that we perceive. I am included in that vast majority. However, it *is* straightforward as a mathematical model and therefore makes perfect sense to a mathematician.

5.3 Missing Mass and Dark Energy

A feature of the Universe that we have not yet considered is the mass that it contains. We can interpret various motions in the universe in terms of gravitational forces, using the concepts of Newtonian mechanics. The Milky Way is rotating and each star moves around the centre of the galaxy under the influence of the gravitational forces of all the other stars — predominantly those closer to the centre than itself. The greater the internal mass the faster it would orbit; for example, if the mass of the Sun were doubled and the Earth was at the same distance, then a year, the time for the Earth to orbit the Sun once, would be only 258 days. Now it is possible to estimate how much mass should be inside the Milky Way stars whose velocities we know (including the Sun) to explain their rates of rotation about the centre of the galaxy. The surprising result of such calculations is that the observed mass of the Milky Way, as judged by the light emitted by the stars within it, accounts for only 10% of the mass required to explain the rotation. This result is confirmed when we look at the motions of galaxies within clusters of galaxies; there seems to be insufficient mass in the galaxies to prevent the clusters from flying apart. Again the observed mass is only about 10% of that required. This has led to the idea of *missing mass*. Nine tenths of the mass in the Universe seems to be in a form that we cannot detect. There are two main theories to explain the missing mass. The first is that there are massive particles (on the scale of the masses of fundamental particles) that interact so feebly with ordinary matter that it is impossible to detect them. These particles have been called WIMPs (Weakly Interacting Massive Particles). The other theory, less favoured, is that it consists of ordinary matter in an invisible form — black holes, black dwarfs, neutron stars and others down to planetary size. By whimsical contrast with WIMPs these are called MACHOs (Massive Astrophysical Compact Halo Objects). So far neither theory has been identified as the total contributor to the missing mass.

A conclusion, based on observations of very distant galaxies, is that the expansion of the Universe may be accelerating, that is to say that the recession speeds of galaxies from each other are increasing with time. This goes completely against what straightforward theory and instinct indicate. Gravity, which attracts galaxies towards one another, should be counteracting the expansion of the Universe and so should be *reducing* their relative speeds with time. If the galaxies *are* accelerating then the total energy of the Universe that we can detect must be increasing so it is posited that there is some source for this extra energy. This has given rise to the idea of *dark energy*, a source of energy that pervades the Universe but which we cannot detect. Nobody has any idea what it could be.

The Universe is indeed weird and wonderful and we shall be examining the main theory that explains its existence. However, before we do that we shall first consider the nature of the matter of which the Universe consists.

Matter and the Universe

Chapter 6

The Nature of Matter

6.1 Atoms and Molecules

I am feeling thirsty so I'll drink some water. That's better — H-two-O is so refreshing. But what is this H-two-O that covers about 70 per cent of the Earth's surface and is so essential to life? Well, we can find out by doing the experiment shown in Figure 6.1. The experiment starts with both tubes full of water — the water columns are supported by atmospheric pressure. An electric current is passed through the water by connecting a battery to two conducting rods (electrodes) inserted in the water and the gasses that come off the positive terminal (anode) and a negative terminal (cathode) are separately collected in the two tubes placed over the terminals. This process, known as electrolysis, breaks up the water into two components, hydrogen and oxygen. Hydrogen and oxygen are different kinds of *atom*, which cannot, by a chemical process, be broken down into any other kind of atom. Two atoms of hydrogen and one atom of oxygen join together to form a *molecule*, a single unit of the material that we call water. The chemical symbol for hydrogen is H, that for oxygen is O and water is represented by H_2O. A representation of a water molecule is given in Figure 6.2. The two hydrogen atoms are connected to the oxygen atom by *chemical bonds.*

An example of a more complex, but quite common, molecule is ethyl alcohol, the essential component of all alcoholic drinks, written

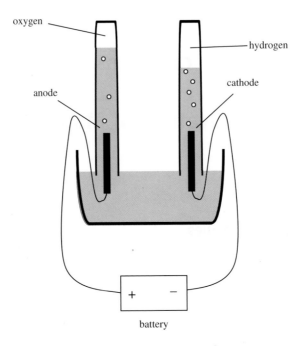

Figure 6.1   The electrolysis of water.

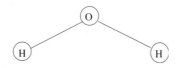

Figure 6.2   A representation of a water molecule.

in the form C_2H_5OH, where C represents a carbon atom, and illustrated in Figure 6.3.

Before the advent of nuclear reactors, in which new types of atom can be created, there were 92 different kinds of naturally-occurring atoms, or *elements*. Some like carbon and iron are well known. Other atoms, like ytterbium, a silvery-metallic element, and selenium, a non-metallic element related chemically to sulphur, are less familiar to most people.

The original idea of an atom originated with the Greek philosopher Democritus (460–357 BC), who wondered what would happen

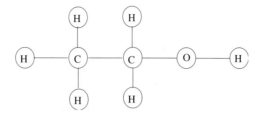

Figure 6.3   A molecule of ethyl alcohol.

if matter was repeatedly divided over and over again. He concluded that eventually one would arrive at an indivisible, indestructible particle of matter. Our word *atom* comes from the Greek, *atomos*, which means indivisible. However, the entities that today we call atoms *can* be divided and we know that they have a substructure of even smaller particles.

6.2 The Discovery of Electrons

The first indication that an atom has components came from the work of the English physicist J.J. Thomson[a] (Figure 6.4) in 1897. He was studying the transmission of electricity through gases contained in a glass tube at very low pressure with an electrode at each end, an experiment that others had previously done. It was known that some kind of radiation came from the cathode — they called it *cathode rays* — but not what the nature of that radiation was. Thomson produced a fine beam of cathode rays and found that they were deflected by both electric and magnetic fields. By passing a beam through a combination of electric and magnetic fields, he demonstrated from the deflection of the beam that cathode rays consisted of negatively charged particles and he was also able to determine the ratio of their electric charge to their mass. It was concluded that these particles, called *electrons*, were derived from the atoms of gas in the tube and hence were constituents of those atoms.

---

[a] Nobel Prize for Physics, 1906.

Figure 6.4   Joseph John Thomson (1856–1940).

A later experiment, in 1910, by an American, R.A. Millikan[b] (Figure 6.5), examined the behaviour of tiny charged droplets of oil when placed in a vertical electric field. The field was adjusted until the droplet was stationary, which was when the upward force due to the electric field, which was proportional to the charge on the droplet, exactly balanced the downward force on the droplet due to gravity. He found that the electric charge on the droplets was always negative and always equal to a multiple of a small charge, which he took to be the charge of an electron. Since the charge of the electron was now known, Thomson's result, which gave the ratio of charge to mass, enabled the mass of the electron to be found, which turned out to be just 1,837 times smaller than that of a hydrogen atom, the lightest atom.

6.3 The Atomic Nucleus

Since atoms are electrically neutral then, if one constituent is negatively-charged electrons, there must be another constituent that is

---

[b] Nobel Prize for Physics, 1923.

Figure 6.5 Robert Millikan (1868–1953).

positively charged. In fact, when an atom loses an electron it is left with most of its mass but a positive charge equal in magnitude to that of the electron; such a particle is known as a positively charged *ion*. At the beginning of the 20th Century, experiments were being done, similar to those done by Thomson with cathode rays, to measure the ratio of charge to mass for various ions. Since their charges were known, this gave a measurement of their masses, and ionic masses turned out to be thousands of times greater than that of the electron.

In 1897 Thomson had proposed what was called the 'plum-pudding' model of an atom, in which the negatively-charged electrons were like the currants in a blob of positive charge containing most of the atomic mass and representing the main bulk of the pudding. Just two years later the New Zealander Ernest Rutherford[c]

---

[c] Nobel Prize for Chemistry, 1908.

Figure 6.6   Ernest Rutherford (1871–1937).

(Figure 6.6), working in Manchester, discovered that an emanation from radium, known as *alpha-rays*, actually consisted of particles with a mass four times that of hydrogen and a positive charge equal in magnitude to twice that of the electron. In 1907 Rutherford asked two people in his laboratory, Hans Geiger and Ernest Marsden, to carry out an experiment in which alpha-particles were fired at a thin gold foil. Most of the alpha-particles went straight through the foil with very little deflection, which is what would be expected from the plum-pudding model, but a few of them were scattered almost back along their original approach direction. Rutherford likened it to a shell from a naval gun bouncing backwards off a sheet of tissue paper. The interpretation of this result was that all the charge of an atom was tightly concentrated into a very small volume, called the *atomic nucleus*, and that the electrons existed in a comparatively large volume around the nucleus. Most of an atom is just empty space through which the alpha-particles could pass unhindered but a few of them

Figure 6.7   James Chadwick (1894–1974).

happened to pass so close to the nucleus that the repulsive force due to the positive charges of the nucleus and alpha-particle (like charges repel each other) gave a large deflection.

The next problem that had to be solved to determine the structure of an atom was that the mass of the nucleus was not proportional to its charge. The first idea to explain this was that the nucleus contained some electrons, so cancelling out some of its positive charge without significantly changing the mass. However, later it was postulated that there were two kinds of particle in the nucleus, one called a *proton* with a positive charge that just balanced the negative charge of the electron, and a particle with no charge, called the *neutron*, the mass of which equalled that of the proton. The existence of the neutron was confirmed in 1932 by James Chadwick[d] (Figure 6.7),

---

[d] Nobel Prize in Physics, 1935.

another Nobel Laureate working in Manchester. There is a light metal called beryllium that is radioactive, the emanation from which is electrically neutral and was thought to be an electromagnetic radiation. Chadwick aimed this emanation at various materials — including paraffin (containing hydrogen), helium and nitrogen — and studied the energies of the recoiling nuclei from the different targets. He was able to show that what came from the beryllium was a neutral particle with the mass of the proton — in fact neutrons.

6.4 The Elusive Neutrino and Antiparticles

Now a complete picture of the structure of an atom had emerged. There was a very tiny nucleus, consisting of protons and neutrons, which contained virtually all the mass of the atom, surrounded by electrons, equal in number to the number of protons in the nucleus, which occupied all the volume of the atom. Figure 6.8 gives representations of atoms of hydrogen, helium and carbon. The diagrams are not to scale. If a nucleus were as big as a human fist then the radius of an atom, containing all the electrons, would be several kilometres. The *atomic mass* of an atom is the sum of the number of protons and neutrons in the nucleus — one for hydrogen, four for helium and twelve for carbon. The *atomic number* of the atom is the number of protons — one for hydrogen, two for helium and six for carbon. It is the atomic number that defines the atom chemically; we shall see later that other forms of carbon exist with different atomic

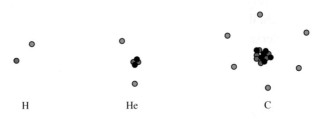

Figure 6.8   Representations of hydrogen (H), helium (He) and carbon (C). Red circles are protons, black circles are neutrons and blue circles are electrons.

Figure 6.9   Wolfgang Pauli (1900–1958).

masses, but they are still carbon because they have six protons in their nuclei. Incidentally, from the description we gave previously of the alpha-particle as having a mass of four hydrogen atoms (approximately equivalent to four proton masses) and a positive charge of two electron units, we can see that an alpha-particle is just the nucleus of a helium atom.

There are some radioactive elements that emit electrons by a process that goes on within the nucleus where a neutron converts into a proton plus electron. The electron shoots out of the nucleus at very high speed — an appreciable fraction of the speed of light — leaving the atom completely, and the nucleus recoils, just as a field gun recoils when it fires a shell. The extent of the recoil and the direction and energy of the electron ejection could be measured and it was found that there was some missing energy and momentum that could not be accounted for. Momentum is a product of mass and velocity that cannot be lost or gained in an isolated system — to which an atom approximates. In 1930 the Austrian-Swiss (later naturalised American) physicist Wolfgang Pauli[e] (Figure 6.9) postulated that the energy

---

[e] Nobel Prize in Physics, 1945.

and momentum were carried off in a particle called a *neutrino* that was produced when the neutron broke down into the proton and electron. This is a strange particle with a very tiny mass (at first it was thought to be without any mass) that barely interacts with matter and is therefore incredibly difficult to detect. Nevertheless, it can be detected with very refined equipment and of its real existence there is no doubt.

A result that comes from high-energy physics experiments is that for every particle there is an *antiparticle*, a particle with the same mass and, in the case of a charged particle, of opposite charge. Thus the *positron* is the antiparticle for the electron — it is essentially an electron with a positive charge and it is output by some radioactive elements when a proton in the nucleus spontaneously converts into a neutron and a positron. Similarly, in some high-energy physics experiments *antiprotons* have been produced, which are essentially protons with a negative charge. In fact, although in the above account we called the particle that came from the disintegration of a neutron a neutrino, it should actually have been called an *antineutrino*. A neutrino does exist and it will be referred to in the following chapter. If a particle and its antiparticle come together the result is the annihilation of the two particles and the production of a great deal of energy in the form of electromagnetic radiation. This is a consequence of the result from Einstein's theory of Special Relativity that shows that mass can be converted into energy. A less obvious result is that in some circumstances it is possible to convert energy into mass! For example, when a high-energy γ-ray photon interacts with a heavy nucleus it is possible for some of the energy of the photon to be converted into a pair of particles — an electron and a positron.

6.5 Quarks

For most purposes, to understand the nature of the matter in our world, and how it behaves, the foregoing description of an atom will suffice. We have seen that the idea that atoms are the ultimate indivisible particles of matter is not true. Can we now say that the proton, neutron, electron and neutrino, together with their antiparticles, are the ultimate particles of matter? The answer is that they are not! When we come to consider how the Universe might have begun we will have to consider conditions that are so extreme that the normal matter that we know now cannot have existed. To help our understanding of these extreme conditions we now describe states of matter that are not parts of everyday life but have been investigated by high-energy physicists with their vast accelerators that speed up charged particles to close to the speed of light and then crash them together in head-on collisions.

Since 1961 there has been developed a new model of the structure of fundamental particles — and there are many other particles than those we have mentioned, particles that can be produced in high-energy physics experiments. This involves a new basic set of particles called *quarks*, of which there are six whose different characteristics are described as their *flavours* and are individually given the names *up* (u), *down* (d), *strange* (s), *charm* (c), *bottom* (b) and *top* (t). The charges associated with quarks are multiples of ⅓ of an electronic charge, either −⅓ or +⅔, and the charges associated with the various quarks, as fractions of an electronic charge, are:

$$u\ \tfrac{2}{3} \qquad d\ -\tfrac{1}{3} \qquad s\ -\tfrac{1}{3} \qquad c\ \tfrac{2}{3} \qquad t\ \tfrac{2}{3} \qquad b\ -\tfrac{1}{3}.$$

There are also *antiquarks* to each of these particles with opposite charges — thus:

$$\bar{u}\ -\tfrac{2}{3} \qquad \bar{d}\ \tfrac{1}{3} \qquad \bar{s}\ \tfrac{1}{3} \qquad \bar{c}\ -\tfrac{2}{3} \qquad \bar{t}\ -\tfrac{2}{3} \qquad \bar{b}\ \tfrac{1}{3}.$$

The electron *is* a fundamental particle in its own right but protons and neutrons are formed by combinations of three quarks of the up-down variety. Thus:

u + u +d has a charge, in electron units, ⅔ + ⅔ − ⅓ = 1 and is a proton.
$\bar{u} + \bar{u} + \bar{d}$ has a charge, in electron units, −⅔ − ⅔ + ⅓ = −1 and is an antiproton.
d + d + u has a charge, in electron units, −⅓ − ⅓ + ⅔ = 0 and is a neutron.
$\bar{d} + \bar{d} + \bar{u}$ has a charge, in electron units, ⅓ + ⅓ − ⅔ = 0 and is an antineutron.

The existence of quarks has not been experimentally verified but a very large number of fundamental particles, not mentioned here, can be explained by combining together, either in pairs or in sets of three, the six quarks and the corresponding antiquarks.

This concludes our description of the nature of matter, sufficient for the purpose of explaining how the Universe and its contents might have originated.

Chapter 7

The Big-Bang Hypothesis

7.1 Origin of the Big-Bang Hypothesis

From Hubble's observations we know that distant galaxies are receding from the Milky Way with a speed proportional to their distance. For example, if a galaxy at a distance of 3,000 million light years has a recession speed of 64,000 kilometres per second then for one at one half of that distance, 1,500 million light years, the recession speed is just 32,000 kilometres per second. Assuming that its recession speed is constant, the distance of a galaxy from the Milky Way at any time in the past can be calculated. At a speed of 64,000 kilometres per second the first of the above-mentioned galaxies has travelled 213 million light years in the last 1,000 million years so its distance from the Milky Way 1,000 million years ago was 3,000 — 213 = 2,787 million light years. Now, this raises an interesting question. How long ago was the distance between the distant galaxy and the Milky Way zero? The answer is about 14,000 million years. Since the speed of recession is proportional to distance, the same answer is found for *any* galaxy, with the conclusion that approximately 14,000 million years ago all galaxies occupied the same space as the Milky Way. Actually, as shown in Figure 4.2, because the distance of a galaxy is increasing with time, it is not necessarily true that it would collide with the Milky Way if its motion were reversed — and hence it is not necessarily true that all galaxies once occupied the same space as the Milky Way 14,000 million years ago. The uncertainty in the observations is sufficiently high for the distance between the Milky Way and the distant galaxy at that time to have been several million light years — close on the scale of the separation of galaxies, but by no means overlapping. That uncertainty also applies to the timescale for closest approach so that 14,000 million years is just an estimate that could be in error by 10 per cent or more.

Despite the uncertainties, we can be sure that, some 14,000 million years ago, the Universe was much smaller and much more congested than it is now. This conclusion has led to the current *Big-Bang Theory* for the origin of the Universe, accepted by most, but not all, astronomers. This theory is that, at some instant in the past, all the energy in the Universe was concentrated at a *point* — a point with no volume that scientists refer to as a *singularity*. That is a challenging idea. The implication of it is that, at the instant the Universe came into being, neither space nor time existed! Again we are in the position that we cannot imagine or understand what this means. Try the following — close your eyes and try to think of nothing — absolutely nothing. You can no more do that than understand — really understand — a Universe of zero volume in which time did not exist. Starting from the singularity, the Universe expanded so creating space and time. Like any sensible person you will ask the question 'What was the state of affairs before the Big Bang?', the answer being 'There is no such thing as *before* the Big Bang because time did not exist until the Big Bang occurred.' You might try again with the question 'Into what did the Universe expand?' to which the answer is that there was no space for the Universe to expand into since the only space that existed was what it created as it expanded.

7.2 What Happened in the Big Bang?

Despite the difficulty — nay, impossibility — of visualising this model, it is possible, nevertheless, to deduce the processes of its development. Starting from the Big Bang, which is not an explosion as its name suggests but rather just an expansion, we can describe what probably happened in certain time intervals. Remember, once the Universe came into being and began to expand, then it *is* possible to talk about time.

From the beginning to 10^{-12} seconds

In this period of the expansion the current laws of physics were not operating. For much of this period it was impossible to distinguish matter and energy or any of the forces that operate in nature. At one stage there ensued a rapid expansion of the Universe, referred to as *inflation*, when the growth speed was faster than the speed of light. By the end of the period there was an incredibly dense region of energetic photons, particles and antiparticles all moving at close to the speed of light. However, from this stage onwards known physical laws operated, thus enabling the processes that occur to be described.

From 10^{-12} to 10^{-10} seconds

The Universe was a seething region of intense radiation with little matter in existence. In the right circumstances energy can change into mass; the example we gave in Section 6.4 involved the production of an electron-positron pair. For photons of the extremely high energies available at this time, the pair production produced quarks and antiquarks. However, although the quark-antiquark pairs were being produced at a high rate they were also being annihilated at a very high rate when they came together.

As the Universe expanded it became cooler and the rate of quark-antiquark pair production fell with time. For some unknown reason, which physics cannot explain, as the rate of production of quark-antiquark pairs fell away, the Universe was left with more quarks than antiquarks. The excess of quarks over antiquarks was tiny, but sufficient to ensure that we live in a Universe made of the stuff we see around us — matter and not antimatter. We know that this must have happened because the protons and neutrons that make up the ordinary matter of the Universe are combinations of quarks, not of antiquarks (Section 6.5).

While the radiation density was high, in the early stages of this period, some combinations of quarks and of antiquarks were giving rise to kinds of exotic particles that can now only be observed in high-energy physics experiments, when streams of protons or heavy-element ions have been accelerated to close to the speed of light and are made to collide head-on.

Time about 10^{-4} seconds

By now quarks ceased to exist as isolated independent particles but they combined together, sometimes in pairs to make particles such as mesons, or in threes to make protons and neutrons. The basic raw materials for the formation of matter in the Universe had come into being, but the Universe was still far too hot for protons and neutrons to bond together to produce atomic nuclei.

Time about 1 second

Isolated neutrons are unstable with an average lifetime of 10 minutes. A neutron decays into three particles — a proton, an electron and an antineutrino (called a neutrino in Section 6.4, but it would have complicated matters to refer to the antiparticle at that time). Protons, on the other hand, are quite stable entities and, left alone, will exist indefinitely. However, a collision of an antineutrino with a proton can give rise to a neutron and a positron (the antiparticle of the electron). Another way of breaking down a proton is to hit it with an electron which produces a neutron and a neutrino. The ways in which neutrons and protons break down are illustrated in Figure 7.1.

To disrupt a proton the energies of the colliding particles have to be very high, which will have been so when the temperature of the Universe was very high, as it was at this stage. Actually, this is a good time to describe temperature from a scientist's viewpoint.

We are all familiar with temperature as it affects our everyday lives. A cold winter day will be 0° Celsius, written as 0° C, a hot summer day can be 30° C, a freezer temperature –18° C and boiling water (by definition) 100° C. The coldest temperature recorded on Earth, at the Russian Antarctic base Vostok in 1983, was –129° C.[a] Physicists,

---

[a] NOAA, http://www.ncdc.noaa.gov/oa/climate/globalextremes.html#lowtemp (Accessed 10/05/12).

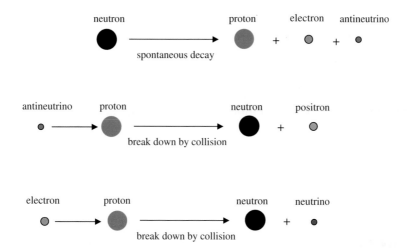

Figure 7.1   The spontaneous breakdown of an isolated neutron and the breakdown of a proton by two processes depending on collisions.

specializing in *low-temperature physics*, vie with each other to produce the lowest temperature possible in their laboratories. Scientifically, temperature is a measure of the energy of motion of the particles that constitute matter when at that temperature. In a gas, for example, the constituent atoms fly around, sometimes bouncing off the wall of the containing vessel, if the gas is contained, and sometimes bouncing off other atoms. If the temperature increases then the atoms move faster and if the temperature decreases then the atoms move more slowly. This raises the question of what the temperature is if the atoms are not moving.[b] It turns out to be approximately –273° C but scientists prefer to define another temperature scale, the *kelvin*, or *absolute*, scale, where this temperature of lowest energy is described as 0 kelvin, or just 0 K (notice, no degree symbol). The temperature intervals on this scale are the same as the Celsius intervals so that the temperature of melting ice (0° C) is 273 K and that of boiling water (100° C) is 373 K. We have used a gas to describe this relationship between particle energy and temperature but the same relationship exists for

---

[b] For theoretical reasons atoms cannot be entirely stationary but they can have a least energy of motion called *zero-point energy*.

atoms bound in a solid. In this case the atoms cannot fly around freely; they vibrate rapidly around some fixed position, like the clapper of an electric bell, only much faster. The picture we now have is that the higher the temperature, the greater is the energy of motion of particles, and temperature on the kelvin scale is proportional to that energy; double the temperature means double the energy.

Time about 100 seconds

Prior to this time, protons and neutrons sometimes managed to combine to form temporary liaisons but, because the temperature was so high, they soon split up again. Another problem that inhibited the formation of associations of protons and neutrons is that the particles passed one another so quickly they did not have the time to combine — bonding needs time to happen. At this stage the temperature had fallen to the point where protons and neutrons could permanently combine to form the nuclei of the light elements helium (He) and lithium (Li), representations of which are shown in Figure 7.2. Most hydrogen has a nucleus which is just a proton but a stable *isotope* of hydrogen, deuterium (D), exists that has a nucleus consisting of one proton plus one neutron. About one six-thousandth of the hydrogen on Earth, in water and in you, is deuterium.

The characteristic of an atom, that defines what element it is, is the number of protons in its nucleus. Since deuterium has one proton in its nucleus, it is just a form of hydrogen. Deuterium was produced at this 100-second stage and this is also shown in Figure 7.2. The amounts of the various light nuclei that formed were governed by the fact that there were about seven times more protons than neutrons in the Universe.

deuterium   helium   lithium

Figure 7.2   Representation of the nuclei of deuterium (1 proton + 1 neutron), helium (2 protons + 2 neutrons) and lithium (3 protons + 4 neutrons).

Time 10,000 years

High-energy radiation had been changing into matter and by this time more of the energy of the Universe was contained in matter than in radiation. Because of cooling, the radiation now had less energy and so was less able to change into further matter. From this time the matter and the radiation in the Universe became independent of each other. As the Universe expanded, creating more space, so the matter spread out and the average density of matter in the Universe fell. The radiation also had a greater space to occupy, the energy per unit volume of the radiation lessened and so the Universe steadily cooled.

Time 500,000 years

Although light-atom *nuclei* had formed in large numbers, the production of whole atoms with a full complement of electrons had not occurred. The temperature had been too high. Electrons are bound to nuclei with a certain binding energy. For this reason, an electron cannot attach itself to a nucleus if it is in a temperature environment such that its energy of motion is higher than its binding energy to the nucleus. By this stage the temperature fell to the level where electrons could attach themselves to nuclei. Electrons attached to the nuclei shown in Figure 7.2 formed deuterium, helium and lithium atoms. However, we recall that the number of protons in the Universe greatly exceeded the number of neutrons and, when electrons attached themselves to free protons, hydrogen atoms were formed (Figure 6.8). This established the pattern that hydrogen is now by far the most abundant element in the Universe.

Starting with a huge amount of energy at a singularity, with no space and no time, we have now progressed to having space and time and also plenty of matter in the Universe. The temperature at this time was about 60,000 K, and falling (it has now reached 2.73 K!), and the created atoms were quite stable. Although the beginning of the Universe is something that physics cannot describe, and our imaginations cannot encompass, an era had now been reached where physical laws can comfortably describe what happens to matter.

However, all the mysteries of the Universe have not gone away — we still have to deal with the concept that we live in an expanding Universe but one that has no boundaries, whatever that means.

The ingredients now existed to create the present structure of the Universe. Next we must consider how these ingredients became the various bodies in the present Universe.

Chapter 8

How Matter Can Clump Together

8.1 Gravitational Instability

After half a million years the Big Bang had provided the material that we now have in the Universe, in the form of light atoms, mostly hydrogen. Later, some of this material was processed in stars to produce heavier elements, whose nuclei contained larger numbers of protons and neutrons, but initially there were only the light atoms hydrogen (including deuterium), helium and a little lithium, well dispersed in gaseous form. We now consider processes by which this diffuse material assembled itself into the compact objects that now populate the Universe.

To understand this, we first describe some work done by James Jeans (Figure 8.1), an eminent British theoretical astronomer, who studied the conditions required to enable a gaseous astronomical body, say a star, to be stable. Stars are complicated objects, with density, temperature and composition varying with distance from their centres, so, following Jeans, we shall deal with a hypothetical spherical gaseous body of uniform density and temperature (Figure 8.2).

Two kinds of force act on the sphere — gravity and gas pressure. Gravity forces pull inwards on all the material of the sphere, just as we are pulled by gravity towards the centre of the Earth. Hence gravity is the force exerted on the sphere by its own mass (which depends on its radius and density) and holds the sphere together. The other force is gas pressure that acts to blow the sphere apart. Gas pressure is the outward force that keeps a balloon inflated. If more gas is forced into

73

Figure 8.1   James Jeans (1877–1946).

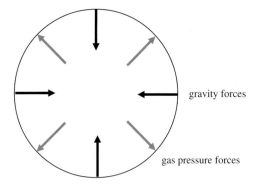

Figure 8.2   The forces acting on a sphere of gas.

the balloon then the density of gas increases, the pressure increases and the balloon expands a little. Another way to expand a balloon is to put it into hot water. Now the gas in the balloon gets hotter, which also increases the pressure within the balloon, so that again the balloon expands. We see that the pressure depends on both the density and temperature of the gas.

It is the balance between gravity and gas pressure that determines whether a gaseous spherical body will hold together or expand outwards. For a sphere of gas at a particular density and temperature, the pressure is independent of the size of the sphere. However, increasing the size of the sphere increases the gravitational force holding it together. For a small sphere the gas pressure force will be dominant and the gaseous sphere will disperse outwards. Increasing the sphere size eventually gives a total mass for which gravity and pressure forces are in balance and at that point the sphere is just stable. Hence for any particular density and temperature there is a minimum mass required for stability, which Jeans found by mathematical analysis, giving a formula in terms of the density and temperature for what is known as the *Jeans critical mass*. This mass also depends on the nature of the material of the sphere which, for our present purpose, is a mixture of hydrogen and helium with a tiny amount of lithium.

Now we consider the ways in which a blob of gas, of density higher than that of its surroundings, could form. There is one mechanism, also explained by Jeans, which can give rise to separation of material into dense blobs starting with more-or-less uniform material. In 1916 Jeans proposed a theory for the origin of the Solar System in which a massive star, passing by the Sun, raised a huge tide on the Sun and pulled a filament of gas out of it. Jeans then showed that this filament would break up along its length into a series of blobs. This is illustrated in Figure 8.3, which shows the behaviour of a filament of gaseous material. The stages are as follows:

(a) A higher-density region somehow forms in a uniform stream of gas. Material near the higher-density region is gravitationally attracted towards it, with closer material being more strongly attracted than that further away.

(b) Material drawn towards the higher-density region leads to the formation of lower-density regions further out. Material outside the lower-density regions now experiences less gravitational attraction inwards than outwards, so tends to move outwards.

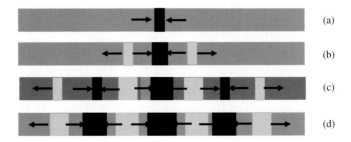

Figure 8.3   Gravitational instability in a gaseous filament.

(c) This outward-moving material creates a new higher-density region that then attracts material just outside it, so producing another lower-density region still further out.

(d) This process continues, producing alternating higher-density and lower-density regions throughout the length of the filament.

If the filament is thick enough then the mass of the individual blobs may exceed the Jeans critical mass, and so give a string of stable condensed gaseous blobs. Analysis of this process gave a formula for the distance between the blobs, which depends on the temperature, density and composition of the gas. This is an example of a process known as *gravitational instability*. A uniform stream of gas is intrinsically unstable and the slightest disturbance in the form of a small density enhancement will trigger a break-up into a string of blobs. Something similar is a matter of common observation; if a water-tap is adjusted to give a very fine uniform stream of water then, often, the stream suddenly breaks up into a line of droplets. This instability is due to a property of water (and other liquids) called *surface tension*.

The stream of gas is unstable, and the question of stability and instability crops up frequently in physical situations. Physical systems tend to move to a state of lowest energy and if by changing itself in some way a system can reduce its energy then it will make that change. We can illustrate the concept of stability with the examples shown in Figure 8.4.

Figure 8.4(a) shows a ball sitting at the bottom of a cup. Its only possible motion is to move upwards, so increasing its energy. Since it

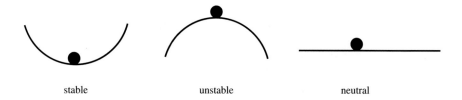

stable   unstable   neutral

Figure 8.4   Three basic types of equilibrium.

needs to have the lowest possible energy it will stay where it is. The ball is in *equilibrium* and, because it will resist any displacement from where it is, it is in *stable equilibrium*. In Fig. 8.4(b) the ball is balanced at the top of an inverted cup. When precisely at the top of the cup then it is in a state of balance, or equilibrium, but the slightest displacement from that position will cause it to continue to fall and, hence, decrease its energy. This is a position of *unstable equilibrium* — the condition of a uniform stream of gas. The slightest increase in density in any part of the stream will trigger off fragmentation into a series of blobs, a state of lower energy. For completeness, we show in Fig. 8.4(c) a ball sitting on a horizontal flat surface. If it moves then it will simply stay in the new position. The energy has neither increased nor decreased, so it has no tendency to prefer one position to another. This ball is in a position of *neutral equilibrium*.

The gravitational instability in the gaseous filament is a one-dimensional phenomenon since the stream breaks up along its length. However, there are also two and three-dimensional occurrences of gravitational instability. Figure 8.5 shows a two-dimensional example, which applies to a thin sheet of gaseous material. In Fig. 8.5(a) a small region of higher density than average is shown. By a process analogous to that for a gaseous filament, the whole sheet breaks up into a number of blobs, as seen in Figure 8.5(b). For a non-uniform sheet, with varying mass per unit area over the sheet, the condensations will neither be equally spaced nor of equal mass, as indicated in the figure.

Three-dimensional gravitational instability of a volume of gas follows the same pattern. If a high-density region forms then, under the

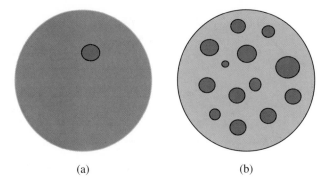

(a)   (b)

Figure 8.5   (a) A thin sheet of gaseous material with a density-enhanced region. (b) Break-up by gravitational instability.

right conditions, the whole volume will break up into a collection of condensations. The term 'under the right conditions' means that if the region of gas was constantly being disturbed then condensations would not have time to form. This relates to yet more of the theory of James Jeans, which gives the time taken for the complete collapse of a blob of gas. The higher the original density of the condensing material, the less time it will take to collapse to high-density — the *free-fall time*. Gas that is constantly disturbed will not be able to form condensations; the initial stage of free-fall collapse is very slow so that before appreciable collapse had taken place the material would be stirred up into a new configuration. It is rather like what happens in lorries that deliver ready-mixed concrete, where the concrete is prevented from separating into its component materials, or setting, by putting it into a rotating drum that keeps it well stirred.

The masses of the condensations formed by gravitational instability are normally about the Jeans critical mass corresponding to the density and temperature of the material in which the condensations form.

8.2 The Role of Turbulence

Another mechanism for producing a high-density blob of material involves a phenomenon known as *turbulence*, often seen in the passage of water along a river. The river may be wide and flowing smoothly and slowly in a flat plain. Then, it enters a mountainous region where, over time, it cuts a narrow ravine. Water entering the ravine increases its speed of flow, and smooth flow in the plain gives way to tumultuous motion in which streams of water crash into one another so that water is thrown high into the air. This is the environment for 'white-water rafting', a popular sport for those looking for thrills.

Many astronomical situations give rise to turbulent motion in which streams of *gas* collide. There is an important difference between the behaviour of a turbulent liquid and that of a turbulent gas, a difference illustrated in Figure 8.6. Water is incompressible — if you squeeze it hard its volume will stay virtually unchanged. Consequently, when two water streams crash head-on, the water can neither change its volume nor can the streams move through each other. The only possible mode of behaviour in the river-flow situation is for water to be thrown into the air. However, when gas streams collide something else can happen. When a gas is squeezed it is readily compressed and, in the turbulent motion of a gas, high-density clumps can be created by colliding streams, as shown in the figure. We shall now describe the possible behaviour of a high density clump.

It is intuitive to suppose that a high-density clump formed in this way would be at a higher pressure than the surrounding material and would just expand until the density, and hence pressure, of its constituent material once again matched that of its surroundings.

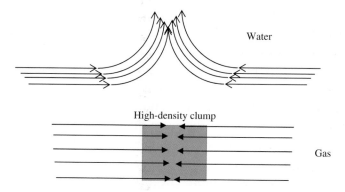

Figure 8.6　The behaviour of streams of water and streams of gas when they collide.

However, this is not necessarily so. If, by some process, the dense clump cools to the level where its mass is greater than the Jeans critical mass then it will further collapse to high density. This depends on a factor that we have not previously considered — the way that the material absorbs heat from, and radiates heat to, its surroundings.

Anyone who has had to inflate a bicycle tyre will know that a compressed gas heats up. The piston of a bicycle pump compresses the air within it until its pressure is high enough to force it through the inner-tube valve. Compressing the air in the pump heats it, and the pump can become quite hot if heavily used. Going back to a compressed clump of gas produced by colliding gas streams, on the face of it the heating would have increased the tendency of the clump of gas to re-expand; it has a higher density and a higher temperature — both of which lead to increased pressure. But that is not the whole story. When all the relevant factors are taken into account, it turns out that the form of development of the compressed region is somewhat counter-intuitive.

8.3 Cooling Processes

The Universe, including the Milky Way galaxy, is traversed by many sources of energy, for example, starlight and cosmic rays, the latter consisting of very high-energy particles. So it was in the early Universe, when the radiation was energetic enough to create matter. The radiation that exists now is absorbed by matter, which is consequently heated. In the absence of any balancing process, the irradiated matter would steadily increase in temperature, without limit. However, there *are* balancing processes. For example, if the gas contains solid material in the form of tiny grains, perhaps only a few microns across, then these would radiate energy, just as a central heating radiator, full of hot water, radiates heat into a room. The higher the temperature, the more it would radiate and if this were the only form of cooling then the gas would heat up until the cooling rate by radiation equalled the heating rate, after which the temperature would remain constant.

There is another, even more effective, process for cooling, which depends on the structure of atoms. The gas would be in the form of atoms, ions (atoms that have had one or more of their electrons removed) and also some free electrons. There is a law in physics, known as the *equipartition principle*, which states that for matter at a certain temperature containing a mixture of different kinds of particle, the average kinetic energy (energy of motion) of each kind of particle is the same. Since the kinetic energy of a particle depends on the product of its mass and the square of its speed, this means that the particles with the least mass move most quickly. Consequently, because of their very low mass, electrons move much faster than the other particles and frequently collide with atoms and ions. These collisions give rise to a number of cooling processes but here we shall explain just one of them.

A branch of modern physics called *quantum mechanics* shows that the electrons in atoms and ions can only exist in states with certain allowed energies. They can jump, or be pushed, from one energy state to another, but what they cannot do is to exist with a non-allowed energy. We now consider the collision of a free electron with an atom, illustrated in Figure 8.7. A helium atom, with two electrons, is used for the illustration. The free electron collides with one of the atomic electrons and pushes into an allowed state of higher energy and the

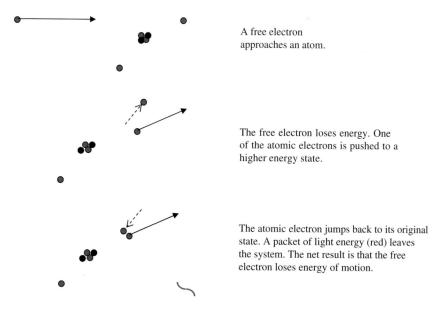

A free electron approaches an atom.

The free electron loses energy. One of the atomic electrons is pushed to a higher energy state.

The atomic electron jumps back to its original state. A packet of light energy (red) leaves the system. The net result is that the free electron loses energy of motion.

Figure 8.7   Cooling due to the collision of a free electron with an atom.

free electron correspondingly loses energy of motion. The atomic electron prefers its original state, with a lower energy, and so the displaced electron spontaneously jumps back to that state. The energy so released is converted into a packet of radiation, a photon (usually visible or ultraviolet light), which travels out of the neighbourhood with the speed of light. The net effect is that the original free electron has lost energy of motion. Due to collisions, this loss of energy is shared with all the other particles, meaning that the average energy of motion in the gas is reduced, i.e. it cools.

There are similar cooling processes involving molecules. All these cooling processes give a greater cooling rate when the material density is higher, since this increases the rate of collisions *per particle*. Another controlling factor is the temperature itself; the higher the temperature, the greater the speed of the electrons and the number of collisions they make, and hence the greater is the cooling rate. Knowledge about the various cooling processes that operate in astronomical contexts goes back a long way. The British atomic physicist, Michael Seaton (1923–2007), gave the theory of cooling by the excitation of atoms and ions in 1955 and Chushiro Hayashi (1920–2010), a Japanese astrophysicist, analysed the role of dust cooling in 1966.

Given these cooling processes we can follow the evolution of a compressed clump of gas in Figure 8.8.

(a) The region is compressed to give a clump with an increase of density and temperature. For simplicity of illustration, the clump is shown as a sphere (circle in projection) but normally it would be of irregular shape.

(b) The pressure in the clump is higher than that in the surrounding gas so the clump begins to expand. However, it is also cooling so the pressure is falling. Cooling is a much faster process than expansion so there is a large fall in temperature during a period of little expansion.

(c) The temperature continues to fall until eventually the pressure in the clump falls below that of its surroundings. Now the region begins to contract due to the squeezing effect of the external pressure.

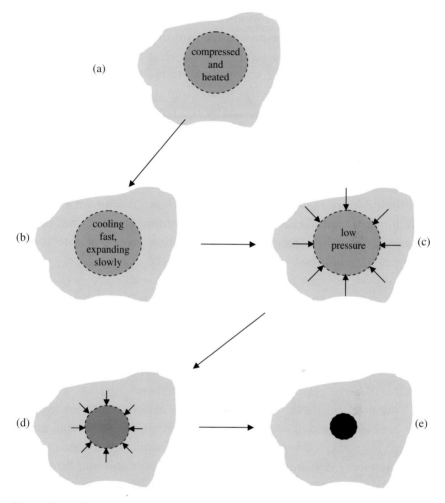

Figure 8.8 The stages in producing a high-density, low-temperature condensation: (a) The clump is compressed and heated by turbulent collision. (b) The clump expands slowly but cools quickly. (c) The pressure falls below that of the surrounding gas and compression sets in. (d) Higher density gives further cooling and further compression. (e) The final equilibrium state.

(d) Compression to higher density increases the cooling rate while the falling temperature reduces the cooling rate. Initially the density effect is stronger and both temperature and pressure fall as the region collapses.

(e) Eventually a state of equilibrium is reached. The clump has a higher density and lower temperature than its surroundings but it has the same pressure so there is no tendency for it either to be compressed further or to expand. At the same time both the compressed region and the surrounding material are in thermal equilibrium with the radiation in the Universe.

To summarise, compressing a gas in an astronomical context can eventually give a final density higher than that when it was first compressed and at a temperature lower than that of its surroundings. The high-density region can be in pressure equilibrium with the surrounding gas and also in thermal equilibrium. Heating due to the absorption of radiation is balanced by cooling processes both in the compressed clump and in the surrounding gas.

However, another factor must be taken into account — gravity. If the mass of compressed gas exceeds the Jeans critical mass then it would collapse further under the influence of gravity. The presence of external material means that the critical mass will not be the same as that calculated by Jeans — his spheres were situated in a vacuum — but the external medium, pressing inwards, will actually increase the rate at which the clump of gas collapsed further.

Two mechanisms by which the gaseous material can form clumps have been described. The first happens spontaneously through gravitational instability of a *quiescent* mass of gas while the second, *turbulence*, requires a non-quiescent state. We shall now consider how these processes give us the range of compact bodies in the Universe, from clusters of galaxies to the small bodies of the Solar System.

Chapter 9

The Universe Develops Structure

The contents of the Universe show a hierarchical structure, consisting of a sequence of entities of ever-decreasing mass, the formation of which is one of the essential topics of this book. This structure is displayed in Figure 9.1.

It is tempting to speculate that large isolated bodies of gaseous material, destined to become superclusters of galaxies, were first produced by one or other of the processes described in Chapter 8. Then, within those condensing clouds, smaller condensing regions formed that eventually became clusters of galaxies. The same kind of process took place, on a smaller and smaller scale, in material that became progressively denser, as one progressed down the sequence of object sizes displayed in Figure 9.1. However, such speculation is wrong — at least for galaxy formation. The mechanism that led to the formation of galaxies, clusters of galaxies and superclusters is still a topic of debate and research and no agreed model has yet emerged.

To form a structured Universe there has to be fragmentation and clumping of its material at some stage. Chapter 8 described two processes by which this could occur. The first is due to gravitational instability that spontaneously breaks up a body of gas into blobs of about a Jeans critical mass for the local conditions. For this to happen the material should be reasonably quiescent, and certainly *not* turbulent; the early stage of free-fall collapse is very slow and the material must not be excessively stirred-up at that stage. The second mechanism requires conditions that *are* turbulent, involving the collision of turbulent

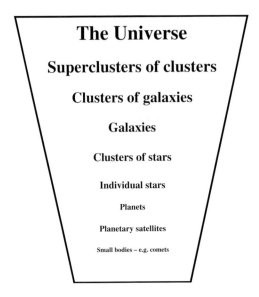

The Universe

Superclusters of clusters

Clusters of galaxies

Galaxies

Clusters of stars

Individual stars

Planets

Planetary satellites

Small bodies – e.g. comets

Figure 9.1   The hierarchical structure of the contents of the Universe.

streams of matter to compress the material and to trigger the process illustrated in Figure 8.8. It is not known whether the early Universe expanded in a smooth uniform way or whether it was turbulent. We shall see that it is possible that its initial fragmentation was by a process other than the two just described. Thus there are several possible means of galaxy formation that need to be explored, based on the various ways that the initial clumping of the Universe could have occurred.

It is not uncommon in the development of scientific ideas that there is no single accepted theory. A theory is advanced, examined, and then, if found wanting, is replaced by another that better agrees with observations. Where more than one theory apparently explains the facts, then recourse can be made to a principle put forward by a 14th Century English Franciscan monk and philosopher, William of Occam (Figure 9.2). This principle, called *Occam's razor*, is given in Latin as *entia non sunt multiplicanda praeter necessitatem*, which literally translates as *entities should not be multiplied beyond necessity*. Applied to scientific theories, the principle is interpreted as 'Given a number of possible theories or explanations, the simplest is to be

Figure 9.2   William of Occam (1285–1349).

preferred'. The reference to 'razor' implies that unnecessary embellishments should be shaved away.

Returning to the expanding Universe, if a clump is formed, by either gravitational instability or turbulent compression, then, from the local density and temperature, we should be able to estimate the masses of those condensations that, by any mode of formation, should be close to a Jeans critical mass. Figure 9.3 shows the estimated average density and temperature of the Universe as a function of the time from the Big Bang. At one million years after the Big Bang the density was 10^{-19} kilograms per cubic metre and the temperature about 1,000 K. This gives a Jeans critical mass of 6×10^{35} kg, equivalent to 300,000 times the mass of the Sun — a substantial mass but less than one hundred-thousandth of the mass of an average galaxy. With the further passage of time, the Jeans critical mass decreased somewhat, the net result of falling density, which increases the Jeans critical mass, and falling temperature, which reduces it. The variation of the Jeans critical mass over the age of the Universe, shown in Figure 9.4, is quite small.

As previously mentioned, for a condensation to form by gravitational instability it must not be disturbed early in the process of condensation. For the density of the Universe at age one million years the free-fall time is over 6 million years so the conditions would need to

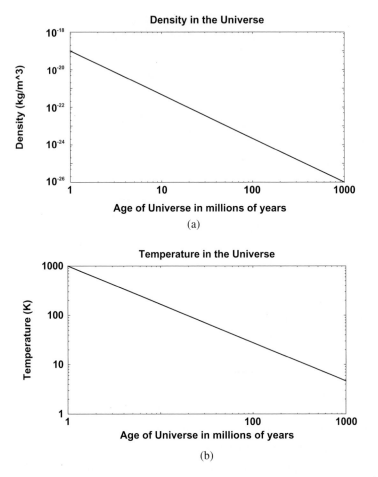

Figure 9.3   The variation of (a) the mean density and (b) the temperature of the Universe with its age.

be reasonably quiet over that period. The older the Universe, the less is its average density and the greater is the free-fall time; at present it is roughly 2×10^{10} years, more than the age of the Universe!

From the above considerations it seems that the first condensations that formed in the expanding Universe had masses a few hundred thousand times that of the Sun — similar to the mass of a globular cluster. In the next chapter we shall see how stars could form in such a condensation.

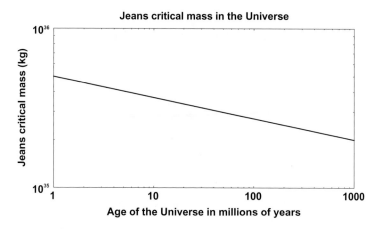

Figure 9.4 The Jeans critical mass for Universe material as a function of the age of the Universe.

If the Universe began with the formation of widely dispersed globular clusters, each containing a few hundred thousand stars, then these clusters must have amalgamated to explain present galaxies and even larger entities. There is a different, and perhaps more promising, idea. It is believed that, after the initial inflationary period following the Big Bang, there may have been lumpiness in the Universe. There is some evidence for this. In 1989 NASA launched the COBE (COsmic Background Explorer) satellite that measured the radiation coming from various directions, which gave corresponding measures of temperature. The results are shown in Figure 9.5 for three different frequencies of radiation. What the measurements showed was that the present mean temperature of the Universe is 2.726 ± 0.010 K but that there were slight variations in different directions, indicating non-uniformity in the large-scale structure of the Universe. These could be the residue of lumpiness in the early Universe and, if so, then these lumps could have been the structures giving individual galaxies. The lumps would move apart due to the overall expansion of the Universe but the lumps themselves, through their self-gravitational forces, would not expand in size at the same proportional rate and so would become increasingly identifiable as separate entities (Figure 9.6). The formation of globular clusters would then occur within each galaxy.

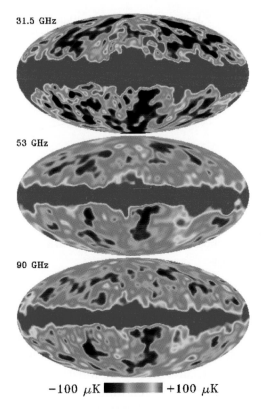

31.5 GHz

53 GHz

90 GHz

$-100 \ \mu K$ ■■■■■ $+100 \ \mu K$

Figure 9.5   Images showing temperature variations with direction at three radiation
frequencies.

Galaxy formation will be considered further in Chapter 11 but,
before that, we shall look at the process of star formation and evolution, which would occur in much the same way whatever the scenario
for galaxy formation.

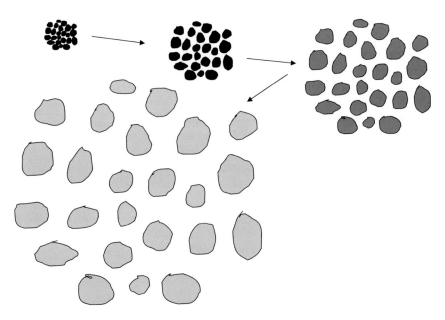

Figure 9.6 A two-dimensional schematic representation of galaxy formation. Overall expansion of the Universe was modified within galactic lumps by gravitational attraction, giving increasing separation of galaxies as the expansion proceeded.

Stars, Stellar Clusters
and Galaxies

Chapter 10

The First Stars are Born, Live and Die

10.1 The Birth of Stars

A typical galactic, or open, cluster, the Pleiades, is shown in Figure 1.3. Galactic clusters normally contain a few hundred stars that are sufficiently separated to be seen individually. By contrast, the globular cluster M13, seen in Figure 1.4, contains several hundred thousand stars and in the central region they cannot easily be resolved. However, there is another important difference between the two kinds of cluster — their component stars differ in material composition.

The lines that appear in stellar spectra (Figure 2.5) reveal the kinds of elements they contain. From the spectra of stars in a globular cluster it is found that they consist almost entirely of original material from the Big Bang, hydrogen and helium. They also contain tiny components of heavier elements, e.g. carbon, oxygen, magnesium, aluminium, silicon, sulphur and iron. These heavier elements were not produced directly in the Big Bang — they were produced by nuclear reactions in stars that transformed the original hydrogen and helium into heavier products. It is believed that the material of globular-cluster stars was once part of earlier stars that started their lives consisting of just the Big-Bang light elements. These early stars synthesised the heavier elements, which were then ejected and mixed with primordial material to produce the material from which, at a later time, the globular-cluster stars formed. These earlier first-generation stars, consisting of pure hydrogen and helium, with perhaps a smattering of lithium, have never been observed but theory suggests that they should once have existed.

The heavier elements in stars include some metals, and astronomers describe the extent to which stars contain heavier elements as their *metallicity*, although many elements contributing to metallicity, e.g. carbon and oxygen, are not metals. Stars in galactic clusters have up to 2% of their mass as heavier elements; they are said to have high metallicity and are referred to as Population I stars. Stars in globular clusters contain a much lower proportion of heavier elements; they are said to have a low metallicity and are called Population II stars. The as yet unobserved but postulated stars with zero metallicity are labelled Population III stars. Within Population I and Population II stars there are varying degrees of metallicity, representing different extents of nuclear processing of their constituent material.

We now consider how Population III stars may have formed from the original Big-Bang material. In Section 8.1 a process was described whereby a Jeans critical mass of gas could spontaneously separate itself from the surrounding gas and begin to collapse. Although in Chapter 9 we found that the masses of such spontaneous condensations in the early Universe were much too small to form galaxies, that does not mean that they did not form — indeed, they were probably the nurseries within which the first Population III stars were born.

Free-fall collapse starts off very slowly, almost imperceptibly, gradually accelerates and then, in the final stages, it is extremely rapid. In Section 8.2 it was explained that when flow becomes rapid the motion will tend to become turbulent and that collisions of turbulent elements can then lead to the formation of high-density, high-temperature regions. Because of grain radiation and atomic and molecular cooling processes, a high-density region cools quickly before it appreciably expands and, if its mass then exceeds the Jeans critical mass, it will collapse to form a high density condensation within the original larger, more slowly collapsing, body of gas. This higher density region will itself begin to collapse, slowly at first but eventually generating turbulence, so there could be a hierarchy of condensations, successively denser and cooler than the material from which they were derived.

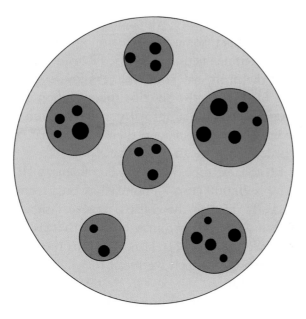

Figure 10.1   An hierarchical system of condensations within a collapsing cloud of primitive Universe material.

This system of condensations is schematically illustrated in Figure 10.1. From Figure 9.4 it seems likely that the original condensations were of globular-cluster mass; at the final level of the hierarchical system the condensations produced will be of stellar mass.

This process of forming the first stars could only have happened early in the expansion of the Universe. If the mean density of the Universe became so low that the free-fall time was almost the same as the current age of the Universe, then forming early stars in this way would clearly be impossible.

This process describes how Population III stars, consisting of primordial Universe material, could have started to form. At the stage where they are beginning their collapse to become stars like the Sun, they have a density considerably greater than the average density of the Universe, but still very low by normal everyday standards. Such bodies of gas, starting the process of collapse that will eventually lead to them becoming stars, are called *protostars*. At the time a gas condensation becomes a protostar, its density would typically be of order

10^{-14} kilograms per cubic metre, which, for a solar-mass, gives a radius of 3.6×10^{11} kilometres, some 2,400 astronomical units. Because of the cooling that accompanied the increased density (Section 8.3), the temperature of the protostar would be very low, somewhere in the range of 10 to 50 K. We now describe the evolution of a solar-mass protostar towards becoming a normal star like the Sun, as originally described by Chushiro Hayashi, and then beyond that stage to what could be regarded as the death of the star.

The free-fall collapse time of a protostar of density 10^{-14} kilograms per cubic metre is 20,000 years — quite short on a cosmic timescale. Although the early stage in the collapse was very slow, heat energy would have been generated within it because the gas is being compressed (the bicycle-pump effect). However, the protostar is so diffuse that it is *transparent*, by which we mean that light, or indeed most electromagnetic radiation, can pass freely through it. If we were to look at the sky through such a protostar, the stars would be seen shining brightly and we would hardly notice that the protostar was there. For this reason the heat generated at this early stage of collapse was simply radiated away and the protostar temperature barely changed. This state of affairs lasted for a considerable time — in fact, most of the free-fall collapse time. This is the blue stage in Figure 10.2.

As the collapse speed increased, the protostar became smaller and denser and, importantly, more opaque to radiation. Heat was being produced at an ever greater rate, and was less able to escape, so that the temperature of the cloud rose at an accelerating rate. By the end of this stage — the light orange region in Figure 10.2 — the surface temperature of the cloud increased to about 100 K. Eventually the collapse became extremely rapid with the surface temperature increasing to several thousand K and the interior temperature much higher than that. The interior pressure built up, opposing the force of gravity, until eventually the pressure and gravity forces were in approximate balance and the rapid collapse ceased. The protostar had developed into a *young stellar object* (YSO), which was in a state of equilibrium, or nearly so. Since it was in equilibrium, its mass was the Jeans critical mass although, since the density and temperature are not uniform throughout the star, a theoretical derivation of the critical mass would

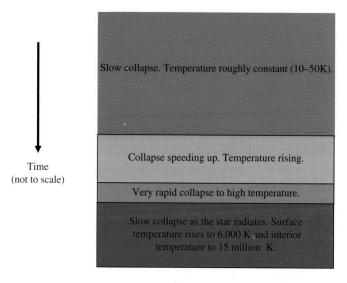

Time
(not to scale)

Slow collapse. Temperature roughly constant (10–50K).

Collapse speeding up. Temperature rising.

Very rapid collapse to high temperature.

Slow collapse as the star radiates. Surface temperature rises to 6,000 K and interior temperature to 15 million K.

Figure 10.2 Stages in the collapse of a protostar.

be quite complicated. The young star has now reached the bottom of the dark orange region of Figure 10.2.

The star was now a hot, radiating, luminous ball of gas. It was losing energy — that which it radiated away — and its reaction to this was to collapse slowly and, paradoxically, to become *hotter*. The energy released by the collapse provided both the energy that was radiated away *and* the increase in thermal energy required to heat up the star. For a solar-mass star, during this stage the surface temperature increased from about 3,000 K to nearly 6,000 K with the interior temperature increasing to about 15 million K. At this temperature the hydrogen in the core of the star began to undergo nuclear reactions that generated large amounts of energy and the star reached the long-lasting main-sequence stage of its development. This takes us to the bottom of the red region of Figure 10.2.

10.2 The Life of Stars

The innermost regions of a newly-formed main-sequence star contain a hydrogen-helium mixture at very high density and temperature and a chain of nuclear reactions occurs, the net result of which is that four atoms of hydrogen are converted into one atom of helium with some by-products. This reaction chain is shown in schematic form in Figure 10.3.

In the first stage of the chain, two protons (hydrogen-atom nuclei) combine to give a deuterium nucleus plus a positron and neutrino. Remember, deuterium, an isotope of hydrogen, has one proton and one neutron in its nucleus. Each deuterium nucleus then combines with a proton to give a helium-3 nucleus. The abundant isotope of helium is helium-4, containing two protons and two neutrons in its nucleus. Helium-3 with one less neutron is stable and constitutes about 0.013% of normal helium gas. In the final stage, two helium-3 nuclei combine together to give a helium-4 nucleus plus two protons. The net effect of the three processes, involving five reactions since the first two occur twice, is that four protons give a helium-4 nucleus, two positrons and two neutrinos.

The final products of this chain of reactions have slightly less combined mass than the four protons that were consumed by it. This lost mass is converted into energy, satisfying Einstein's equation $E = mc^2$,

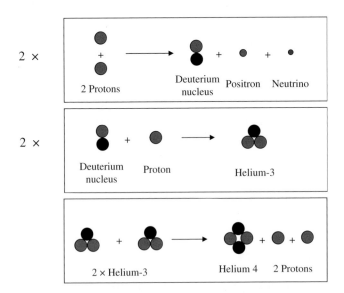

Figure 10.3   The stages in the conversion of hydrogen to helium.

and this energy is emitted by the star in the form of electromagnetic radiation. In the last stage illustrated in Figure 10.2, the energy radiated by the star, and the energy for increasing its temperature, was provided by the gravitational energy released by the collapse of the star. With nuclear energy available, the star ceases to collapse and can still emit radiation. During this main-sequence stage the star's brightness and size remain approximately constant. The Sun became a main sequence star 4,600 million years ago and will remain on the main sequence for the next 5,000 million years. It will leave the main sequence when hydrogen becomes exhausted in its central core.

10.3 The Final Journey

Since nuclear reactions proceed most rapidly at the centre of the star, where the temperature is highest, that is where hydrogen first becomes depleted. When this happens, less energy is produced in the core, the pressure that opposes gravitational forces is reduced and the core begins to contract. The gravitational energy so released heats up a shell of material, still hydrogen rich, surrounding the hydrogen-depleted core. Nuclear reactions continue in this shell, which gradually expands as the hydrogen reactions spread outwards. This state of *hydrogen-shell burning* is illustrated in Figure 10.4.

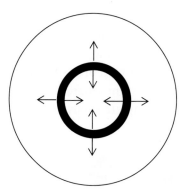

Figure 10.4 Hydrogen-shell burning. Pressure forces, acting in the directions shown by the arrows, compress the core (together with gravity) and expand the outer parts of the star to produce a red giant.

The centre of the star now collapses under the combined effect of gravity and the pressure exerted by the shell burning, which pushes both inwards and outwards. The outward push causes the star to expand, at the same time reducing its surface temperature. In this condition, where the radius of the star becomes very large and its surface temperature is much reduced, the star has become a *red giant* with radius about one astronomical unit.

A safety-valve mechanism ensures a steady and controlled compression of the core that, at this stage, is virtually all helium. As the core becomes denser, so its temperature increases, and the pressure rises to oppose and slow down further collapse. However, when the temperature reaches 100 million K, a new phenomenon occurs — nuclear reactions involving helium. The reactions take place in two stages with the net result that three helium nuclei combine to form a carbon nucleus (Figure 10.5). Since a helium nucleus is an alpha-particle, this is known as the *triple-alpha reaction*. It is a new and powerful source of energy which grows rapidly — reactions give a higher temperature and higher temperature gives an increasing reaction rate.

The new source of energy in the core produces a situation similar to that existing at the main-sequence stage. The star shrinks again, taking up a configuration different from, but similar to, that of the main sequence. Eventually fuel depletion in the core occurs again, but this time it is helium fuel running out. The core is now mostly carbon, and helium-shell burning is established. This again compresses the core and expands the star towards a red-giant configuration. The energy production is now so high that during the final stages of helium-shell burning the outward pressure ejects outer material of the star. This ejected material is illuminated by the star and, although it consists of one or more complete shells, it is seen as rings surrounding

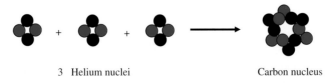

3 Helium nuclei   Carbon nucleus

Figure 10.5   The net result of the triple-alpha reaction.

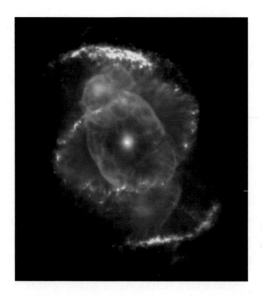

Figure 10.6 The Cat's Eye Nebula.

the star. An example of this phenomenon, known as a *planetary neb-ula* (it has nothing to do with planets!), is shown in Figure 10.6 where several shells of material have been ejected. This nebula, the Cat's Eye Nebula, featured on a UK postage stamp in 2007, which commemorated the 50th anniversary of the popular BBC television series *The Sky at Night*, hosted by Patrick Moore.

The development of the star following helium-shell burning depends on the star's mass. For a star of solar mass or less, all the outer layers of the star are stripped off, leaving the core in the form of a *white dwarf*, a small body of very high density in which nuclear reactions no longer occur. A white dwarf shines because of the energy stored within it. It becomes dimmer with time until, eventually, it becomes invisible, at which stage it is a *black dwarf*.

The material of a white dwarf is the highly-compressed core of the star, mostly carbon, and it has a state very unlike that of normal material. White-dwarf material is in a *degenerate state*; a white dwarf with the mass of the Sun is about the size of the Earth. To get a feeling for this, a teaspoonful of degenerate matter from a white dwarf would have a mass of 10 tonnes! Actually, although it was not previously mentioned, during the development of the star towards the white-dwarf stage, the core was sometimes in a degenerate state, which affected some detailed aspects of the star's evolution.

Stars of greater mass, which run through their evolutionary paths more quickly, do not end up as white dwarfs. The temperatures generated within them are so great that further nuclear reactions can take place after the triple-alpha reaction. Some of these, all involving nuclei, are:

Helium + Carbon	$\Rightarrow$	Oxygen
Carbon + Carbon	$\Rightarrow$	Neon + Helium
Carbon + Carbon	$\Rightarrow$	Sodium + proton
Carbon + Carbon	$\Rightarrow$	Magnesium + neutron
Oxygen + Oxygen	$\Rightarrow$	Silicon + Helium

Because of these extra reactions, a massive star will go through several stages of shell burning with alternating stages of expansion and contraction. However, the process of building heavier elements stops with the formation of iron. For reactions up to those producing iron, the reaction products have less mass than the reacting nuclei and the difference of mass appears as energy. This heats up the system and promotes new reactions. For nuclear reactions that produce elements heavier than iron, the products have a *greater* mass than the original reacting nuclei. For such reactions to occur energy must be supplied from somewhere to create the required extra mass. This energy cannot come from the star since that would cool down the system and hence *prevent* further reactions from occurring. The shell structure of a massive star in its final stages of development is shown in Figure 10.7.

When the iron core grows to a certain size, the pressure within it becomes so great that the material becomes degenerate. However, at an extremely high pressure the iron atoms cannot retain their normal atomic structure, with a core of protons and neutrons surrounded by electrons. The protons and electrons are squeezed together and combine to become neutrons so that the core becomes a solid mass of tightly-packed neutrons. The process of neutron formation happens very quickly and as the core rapidly shrinks so outer material rushes in to occupy the released space. This material bounces off the neutron

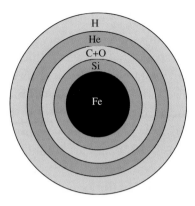

Figure 10.7 Composition shells in a highly evolved massive star. H = hydrogen; He = helium; C = carbon; O = oxygen; Si = silicon; Fe = iron.

Figure 10.8 The Crab Nebula.

core and then reacts violently with material that is still moving inwards. The result is an explosion that shatters the star and expels all the material outside the neutron core. The explosive event is a *super-nova*; the debris from a supernova, the Crab Nebula, first observed in 1054 AD, is shown in Figure 10.8.

The neutron-core residue of the supernova becomes a *neutron star*, with mass up to three solar masses but a diameter of only

10 kilometres. A teaspoonful of neutron-star material would have a mass of 5,000 million tonnes! Neutron stars rotate rapidly and as they do so they emit fine beams of radio waves, which trace out the surface of a cone as the neutron star spins. If these intersect the Earth they are recorded as regular radio pulses — sometimes at a rate of 1,000 pulses per second. Such sources are *pulsars*; at the heart of the Crab Nebula is the Crab Pulsar. When pulsars were first detected by Jocelyn Bell Burnell (b. 1943) in 1967 it was thought that these extremely regular pulses might be some form of communication from an extraterrestrial civilisation — or, as the popular press reported, messages from 'little green men'.

Supernova explosions have an important effect on the composition of the Universe. Some of the released energy fuels reactions that produce elements heavier than iron. All the heavier elements produced by the star — up to iron before the supernova stage and some heavier than iron due to the supernova — are released into the Universe and mix with primordial hydrogen and helium, creating the material that forms Population II stars in globular clusters. Given the time since Population III stars were formed they may all have gone through their life-cycles and no longer exist. However, it is just possible that some of the least massive stars, which evolve most slowly, might still be around, and astronomers are always hopeful that a Population III star might, one day, be detected.

The death of a star increases the inventory of heavy elements in the Universe so that as time progresses the metallicity of stars tend to increase. The material of a star in a galactic cluster with 2% of its content as heavier elements has probably been processed by supernovae in several previous stars.

Another aspect of the death of a star should be mentioned. For a neutron star, with mass more than three solar masses, the pressure at its centre would be too great even for the compressed neutrons to resist. In that case a *black hole* would form. The star would collapse without limit and eventually form a body of finite mass but, theoretically at any rate, shrunk to a point. No radiation could escape from such a body — hence its name — and it could only be detected by its gravitational effects. The existence of black holes was predicted long ago by Pierre Laplace (1749–1827), a prediction based on the concept of the *escape speed*. The greater the speed a ball is thrown upwards, the higher it will go before it starts to fall. If a ball, or rocket, is sent up with a speed of more than 11 kilometres per second then it will escape from the Earth's gravity and it will never return. Laplace argued that light could not escape from a body if the escape speed from that body was more than the speed of light, a way of saying that the body would be a black hole.

The description given here for the birth, life and death of Population III stars also applies to Population II and Population I stars. The small component of heavier elements in these later stars would not substantially affect the evolutionary processes.

Chapter 11

The Formation of Globular Clusters and Galaxies

11.1 What Constitutes the Missing Mass?

When considering how galaxies and their contents form, it is important to take into account that the matter we can either see, such as stars, or can infer from observations, such as interstellar material, accounts for only one tenth of the mass of the Universe. Unless the mass that we cannot see — dark matter — really exists then we can explain neither the rotation rates of galaxies nor the stability of galactic clusters. The two main candidates being considered for dark matter are:

WIMPs (Weakly Interacting Massive Particles). These are elementary particles, abundantly produced by the Big Bang, of which a sufficient number have survived to account for the missing mass. They interact with matter so feebly that, as yet, no means of detecting them has been found. If they exist, then the only indication of their presence is their gravitational effect.

MACHOs (Massive Compact Halo Objects). These objects consist of ordinary matter that emit either no radiation or, perhaps, too little radiation to be detected. Among the candidates that have been suggested for these objects are black holes, black dwarfs (faded white dwarfs), brown dwarfs (bodies intermediate in mass between a planet and a star) and smaller, planetary-size bodies.

Experiments to detect dark matter are in progress in various parts of the world. In one project, UKDMC (UK Dark Matter Collaboration), detectors have been placed in the deepest mine in Europe, Boulby mine in North Yorkshire with a depth of 1,100 metres, a level to which cosmic rays cannot penetrate, so excluding their effects on the detectors. Some activation of the detectors has been recorded but, so far, the evidence for WIMPs is not strong enough to be accepted by the scientific community.

Experiments to detect MACHOs have depended on a result coming from Einstein's Theory of General Relativity, which predicts that a ray of light passing close to a massive object will be deflected. Indeed, it was the observation of this particular phenomenon that led to the widespread acceptance of General Relativity Theory. The deflection of light is also predicted by Newtonian mechanics on the basis that light can be regarded as particles moving with the speed of light. However, General Relativity predicts twice the deflection that would come just from the action of Newtonian mechanics. Einstein published his theory in 1915 when he was in Berlin and World War I was in its second year. The opportunity to test Einstein's prediction occurred at the time of a total solar eclipse in 1919. With direct light from the Sun obscured by the Moon, it is possible to see the light from stars which passed very close to the Sun's surface. Because of the deflection of the light due to the Sun's gravitational field, the observed star directions were slightly displaced from where they were normally seen and, from the displacements, the deflection of the light could be determined. These observations were made by two teams of British astronomers in South America and West Africa, one led by Arthur Eddington (1882–1944) and the other by the Astronomer Royal of the time, Frank Dyson (1868–1939). Their results confirmed Einstein's prediction and General Relativity Theory was soon widely accepted.

The deflection of light when it passes close to a massive object leads to a phenomenon known as *gravitational lensing*. If a massive object, say a brown dwarf, crosses the line of sight to a star then the brightness of the star temporarily increases. The way that this works is illustrated in Figure 11.1. The evidence from the use of this technique

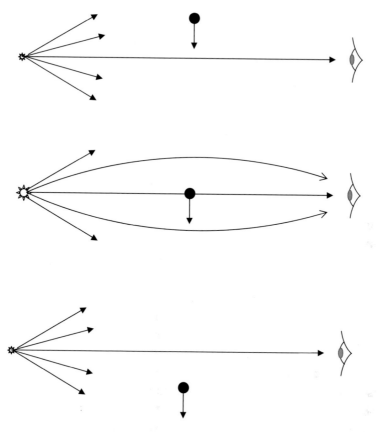

Figure 11.1 The effect of gravitational lensing. When the massive object is on, or close to, the line of sight, more of the light coming from the star enters the eye so that the star brightens.

so far suggests that a large fraction of the missing mass, about 50%, can be ascribed to the existence of MACHOs, but that still leaves the remainder to be accounted for. An important consideration is that, in the early Universe, the matter that now comprises MACHOs was in a form that could contribute to star formation — unlike WIMPs, a form of exotic matter that could not become part of a compact body consisting of ordinary material. With not all of the missing matter accounted for by MACHO detection, the hunt for the existence of WIMPs continues.

Before discussing galaxy formation, we consider some general aspects of star formation and evolution that would apply in any scenario. At an age of one million years, the Universe had an average density of 10^{-19} kilograms per cubic metre and a temperature of 1,000 K. If the Universe were lumpy then the density within a 'lump' would have been a few times the average but it would not affect the general pattern of star formation now described. From either gravitational instability or turbulent compression, clumps of material, roughly of Jeans critical mass, would begin to form. This would be 6×10^{35} kilograms for the initial state of the collapsing material, with a free-fall collapse time of about six million years. Hierarchical break-up of the clump, as shown in Figure 10.1, would eventually lead to stellar-mass condensations that would then undergo the evolutionary development described in Chapter 10. The time taken for a star to go through its life cycle is mass dependent. Table 11.1 shows the time for a star to reach the main sequence — that is, to reach the bottom of Figure 10.2 — and also its lifetime on the main sequence.

Stars that go through their life cycles to give either a white dwarf, which eventually becomes a black dwarf, a neutron star, not detected as a pulsar, or a black hole, all contribute to the missing mass. Once material has converted into one or other of these types of body then it is no longer available for further star formation. For this reason it is inevitable that some of the missing mass *must be* in the form of MACHOs, the only question being whether they could provide the whole of the missing mass or just part of it.

Table 11.1   Times for stages of a star's evolution.

Mass of star (Sun units)	Time to reach main sequence (million years)	Time on the main sequence (million years)
15	0.06	1
9	0.15	30
3	2.5	400
1.5	18	4,000
1	50	9,000
0.5	150	60,000

Figure 11.2 A collision of galaxies (NASA, Hubble).

11.2 How Do Galaxies Form?

The evolution of massive Population III stars gave the production of heavy elements up to iron and the final supernovae provided the energy for the formation of elements beyond iron. This material, ejected into the local environment, became part of the low-metallicity Population II stars that constitute the present globular clusters.

From Table 11.1 it can be deduced that present-day globular clusters formed very early in the life of the Universe. The maximum mass observed for a main-sequence star in a globular cluster is about 0.8 solar masses. This suggests that all stars more massive than this limit in the globular cluster have evolved beyond the main sequence. Table 11.1 indicates that the time for a star of 0.8 solar masses to complete its life cycle is very close to the age of the Universe. From this we may infer that the origin of globular clusters is either contemporaneous with, or closely follows — i.e. was within a few hundred million years from — the formation of the initial Population III clusters. We have already remarked on the fact that no Population III stars have ever been observed and the combination of this fact and the age of globular clusters suggest that the masses of the initial Population III stars probably excluded the smaller mass range. If the least massive Population III stars were, say, of 3 solar masses, seen from Table 11.1 to have a lifetime of 400 million years, then this would explain both how material was available early on for the formation of globular clusters and why it is that Population III stars are not observed today. The lower limit of 3 solar masses is just a suggestion — a lesser value could still explain the observations.

As previously stated, the formation of galaxies is a subject of speculation and uncertainty. Figure 9.4 shows that the Jeans critical mass in the Universe has never been large enough for gravitational instability to give spontaneous galaxy-mass clumps. If the Universe became lumpy after the inflationary period, with the lumps of galactic mass, then galaxy formation is straightforward, with the processes of forming globular clusters and all categories of stars contained within a single lump. Alternatively, if the lumps were smaller, perhaps the size of the very smallest galaxies, then producing larger galaxies would require them to amalgamate in some way. This is the favoured theory at present, that larger galaxies were produced by the amalgamation of smaller ones. There is evidence that galaxies can collide and combine; an example of two galaxies colliding is shown in Figure 11.2 and many other such examples have been recorded. Until recently the furthest, and hence oldest, galaxies that had been observed were formed some three billion years after the Big Bang but in 2004 astronomers in California claimed to have found a galaxy formed less than one billion years after the Big Bang. It is perhaps significant that this galaxy was rather tiny, with a diameter less than 3% that of the Milky Way. If all galaxies formed early on were small then the larger ones could only have formed by amalgamation.

The main types of galaxy are *elliptical* and *spiral*. A typical elliptical galaxy, M32 (Figure 11.3), is an elliptical blob of stars with no discernable internal structure. They have an enormous range of size and content. The smallest are a few thousand light years in mean diameter and contain a few million stars; the largest can be a million light years in mean diameter and contain a trillion (million million) stars — much bigger and more populous than the Milky Way.

A typical spiral galaxy is NGC 6744, shown in Figure 1.5, which resembles the Milky Way. Spiral galaxies have a great deal of structure; the plan view seen in Figure 1.5 shows a strong concentration of stars in the central region and spiral arms coming out from the centre. However, to see other features of the structure we need to look at a side view, as given by the galaxy NGC 4565 (Figure 11.4). The strong

Figure 11.3   M32.

Figure 11.4   The spiral galaxy NGC 4565.

central concentration of stars is in the form of a flattened sphere. Within the central bulge most stars are of the Population II variety. It is considered likely that within the bulge at the very centre of the Milky Way there is a giant black hole. We also see in Figure 11.4 that the part of the galaxy containing the spiral arms is in the form of a disk. The disk contains mostly Population I stars, those most recently formed, and star formation is continuously going on within the disk. The component of the galaxy that is *not* seen, because it is too diffuse, is the galactic halo, a spherical region of diameter so large that it encompasses the whole of the visible part of the galaxy. The stars within the halo are solely older Population II stars and many of these are within globular clusters. The most important component of the halo is what we cannot see — dark matter. Most dark matter in the galaxy is located within the halo — hence the H in MACHO.

However galaxies come about, it is clear that galactic (open) clusters only form once a galaxy exists. The galactic plane is rich in stars and the mean density within the galactic plane is much higher than in the halo. It is the higher density of material in the galactic plane that leads to a lower Jeans critical mass and hence less massive condensations, corresponding to the mass of a galactic cluster containing a few hundred stars. It is also worth noting that the mean density of the interstellar medium, the material between the stars in our galaxy, is about 10^{-21} kilograms per cubic metre, one hundred thousand times greater than the mean density of the Universe. The stars in the galactic plane are a source of occasional supernovae, the blast waves from which can trigger the formation of a cool, dense cloud (Figure 8.8), eventually leading to further star formation. The Sun is a typical Population I star, like those in galactic clusters, and next we describe how such stars are formed.

Chapter 12

Making the Sun — and Similar Stars

12.1 The Ingredients For Star Formation

The solar spectrum (Figure 2.5) is rich in Fraunhofer lines that indicate the presence of many different elements in the outer atmosphere of the Sun. From the number and the strengths of these lines the proportions by mass of the most abundant elements in the Sun have been found, and are given in Table 12.1.

The Sun is a typical Population I star with about 2% of its mass as heavier elements. It is also typical of the stars that are found in galactic clusters. These clusters, usually consisting of a few hundred stars, contain only Population I stars and they occur only in the disk region of the galaxy. By contrast, globular clusters can occur anywhere, but mostly populate the halo region. While the Sun is a field star, moving alone through the galaxy, it is believed that the Sun and other Population I field stars and isolated binary systems began as members of galactic clusters from which they escaped. In considering how galactic clusters form we are almost certainly considering the formation of all Population I stars.

While the postulated Population III stars, and the Population II stars that occur in globular clusters, are old stars formed billions of years ago, the formation of galactic clusters, containing Population I stars, is happening now. Before establishing where and how this happens we should first recall the conditions within our galaxy, briefly described in Chapter 1. The bulk of the galaxy by volume is the interstellar medium, the ISM. The ISM consists of hydrogen and helium,

Table 12.1 The major heavier elements in the Sun.

Element	Proportion by mass (%)
Oxygen	0.97
Carbon	0.40
Iron	0.14
Silicon	0.099
Nitrogen	0.096
Magnesium	0.076
Neon	0.058
Sulphur	0.040

with traces of other gaseous material, together with dust at the 1–2% level by mass. The overall density of the ISM is of the order 10^{-21} kilograms per cubic metre and its temperature is about 10,000 K. What would happen to an astronaut — in his space-suit of course — if he were adrift in such an environment? Would he be burnt to a cinder? The answer is that he would not — he would actually need a heating source in his space-suit to stay alive for any length of time. The temperature of a substance specifies the mean energy of motion of its constituent atoms. Thus a temperature of 10,000 K for hydrogen means that the average speed of motion of a hydrogen atom is about 15 kilometres per second. However, these atoms very sparsely occupy the ISM and a man, with surface area about 2 square metres, would be struck by about 22,000 million of them per second. The rate of delivery of heat by those atoms is such that it would take 3,000 years for them to heat one cubic centimetre of water from freezing to boiling point! Our astronaut is in no real danger from that source.

12.2 Forming Dark, Cool Clouds

Examining the sky with a telescope shows a rich panorama of stars filling the field of view, and with a high quality telescope we would also see many distant galaxies. However, in certain directions the view is obscured — there are dark patches in the sky where some opaque obstructions block the light coming from stars. These obstructing

Figure 12.1 The Horsehead Nebula (N.A. Sharp/NOAO/AURA/NSF).

objects are dense clouds in which the dust is concentrated to the extent that light cannot penetrate them. A rather handsome example of such a dark cloud is the Horsehead Nebula (Figure 12.1).

Dark clouds form by the processes illustrated in Figure 8.8. The ISM material is compressed through the action of a supernova that also injects heavier elements into it in the form of dust. Both these effects enhance the cooling of the affected region. This reduces its pressure so that the surrounding gas compresses the region further and so further enhances the rate of cooling. Another effect of compressing the gas is that molecules begin to form, for example, combinations of carbon (C) and oxygen (O) form carbon dioxide (CO_2), and these molecules provide extra mechanisms for cooling. Eventually the condition is reached where the cooling from the higher-density region just balances the rate of heating by cosmic rays and starlight. The dense cloud is now in pressure equilibrium with the surrounding ISM and in thermal equilibrium with the various sources of heating. At this stage the cloud becomes a dense cool cloud (DCC), typically with density 10^{-18} kilograms per cubic metre and temperature somewhere in the range 10–50 K. In astronomy we sometimes describe things in terms that make no sense in the everyday world. The cloud is 'dense' only in the sense that it is 1,000 times denser than the ISM. There are physicists who carry out experiments in enclosures under conditions of 'ultra-high vacuum', so that the material they are investigating will not be contaminated by bombardment with gas atoms. These ultra-high vacuum enclosures have one hundred times the density of a DCC!

Although a DCC is in pressure and temperature equilibrium with its surroundings, if its mass exceeds the Jeans critical mass then it will begin a free-fall collapse. The free-fall time for a cloud with an initial density of 10^{-18} kilograms per cubic metre is over two million years but the actual time for collapse would be somewhat greater. There are two reasons for this. The first is that in the final stages of collapse, when the cloud becomes opaque, released heat energy is retained, so increasing the temperature, and hence pressure, of the cloud material, which would partially counteract the gravitational forces that drive the collapse. The second reason is that, as the collapse speeds up, turbulent motions within the cloud act like an extra source of pressure, again slowing down the collapse.

Turbulence in collapsing DCCs can be both detected and measured. Such measurements depend on a process, happening in the collapsing cloud, that is not completely understood. The term *laser* is an acronym for 'light amplification by the stimulated emission of radiation'. A laser produces an intense parallel beam of light, the process depending on mirrors that reflect light to-and-fro along the same path. The colour (wavelength) of the light depends on the working substance of the laser, e.g. a gas or a crystal such as ruby. Electrically stimulating the working substance pushes an electron to one of the allowed higher-energy states and when the electron returns to its original state it emits light of a characteristic wavelength. The trick in a laser is to persuade all this light to go in one direction — hence the mirrors. Of similar behaviour to a laser is a *maser*, where the *m* stands for *microwave* and which is exactly like a laser except that the wavelength of the radiation is much greater — of order one centimetre. The working substance in this case can be a molecule, such as water or carbon dioxide, which are constituents of DCCs. By an unknown mechanism, maser emission, with wavelengths characteristic of particular kinds of molecule, is observed from DCCs. What is actually observed is a range of wavelengths around the characteristic wavelengths, the shifts in wavelength being due to the Doppler Effect. By measuring these Doppler shifts one finds variations of source speeds up to 50 kilometres per second or more, both towards and away from the Earth, and these are interpreted as the speeds of motion of turbulent elements within the cloud.

12.3 Forming Protostars

Figure 12.2 gives a schematic two-dimensional representation of the motions of material in the various parts of a turbulent cloud. Here and there within the cloud, turbulent elements are colliding and, when they do so, the material involved will be both compressed and heated. The compressed region will cool on a short timescale, well before the compressed region has substantially re-expanded, and so a cool, dense region is the outcome. In this case, starting with a dense cloud, there will be no hierarchical sequence of dense regions but, in the right circumstances, the compressed region will commence a collapse that leads to the formation of a star, as described in Chapter 10.

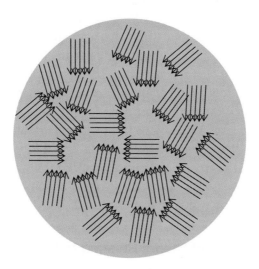

Figure 12.2 A schematic representation of random motions within a generally-collapsing system.

Figure 12.3   The Sun with prominent sunspots.

12.4 The Rotation of Stars

There is a characteristic of main-sequence stars not yet considered but of considerable interest, and that is their rates of spin. The rate at which the Sun spins can be found by observing the motions of sunspots, darker regions on the surface that are the sites of considerable magnetic activity (Figure 12.3). The rate at which the Sun rotates depends upon the latitude — varying from a rotation period of 25 days at the equator to about 34 days near the poles. This is a slow rate of rotation, corresponding to a speed of 2 kilometres per second at the equator, and such low equatorial speeds are characteristic of lower-mass main-sequence stars, in particular those with masses less than about 1.35 times the solar mass. Larger mass stars have equatorial speeds that are much greater and tend to increase with increasing mass although there is a fall-off in speed for the highest mass stars (Figure 12.4). Of course, not all stars of the same mass have the same equatorial speed and what we are referring to is the *average* equatorial speed for stars of different mass. We see that there is a sharp rise in speed for stars of more than about 1.35 solar masses and this must relate in some way to the process, or processes, by which stars are formed.

12.5 Observation and Theory Relating to Star Formation

Main-sequence stars occur with a large mass range, with mass-dependent rotation speeds, and more often than not are members of

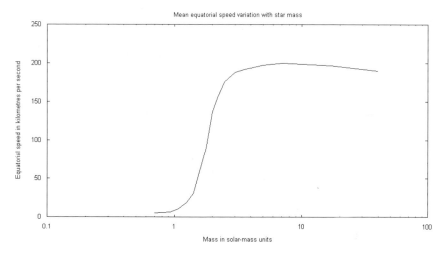

Figure 12.4   The variation of mean equatorial speed with stellar mass for main-sequence stars.

a binary system. It is important to relate theories to as many observations as possible. The observations may not only help to formulate a plausible theory but may also give constraints that can help to distinguish theories that are plausible from those that are not. Since we are concerned with star formation, then observations relating to very young stars are of interest. In 1969 the British astronomers Iwan Williams and William Cremin presented the results of studying a young galactic cluster, NGC 2264, in which they found the masses of the stars and the times of their formation. The main conclusions that could be drawn from their study are:

1. The first stars produced are of mass about 1.4 times the solar mass and thereafter there are two streams of star formation, one producing stars of lesser mass and another producing stars of greater mass (Figure 12.5).
2. The rate at which stars are produced increases with time (for NGC 2264 star formation is still in progress).
3. Greater numbers of stars are produced at lower masses, as is observed for main-sequence stars in general.

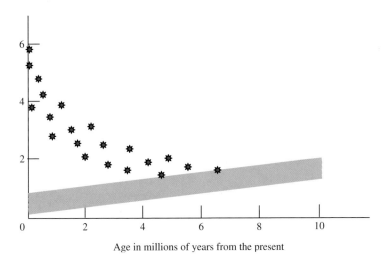

Age in millions of years from the present

Figure 12.5   Schematic representation of the Williams and Cremin results. The grey region gives the formation of lower mass stars and the red symbols give the region of formation of higher mass stars.

In 1979 I published a possible model, tested by computation, which gives results consistent with these observations.[a] As the star-forming DCC collapses, its density increases while, in the early stages, the temperature increase is moderated by radiation out of the partially-transparent cloud. For this reason the Jeans critical mass steadily falls. Another change in the cloud as it collapses is that its turbulence increases, the increase being fed by part of the gravitational energy released by the collapse. When the turbulent speed has increased to the level where colliding turbulent elements sufficiently compress material, the process of star formation begins and the model indicates that this occurs when the stars produced have a mass about 1.35 solar masses. As time progresses so the falling Jeans critical mass enables less massive stars to be produced. Again, with increasing turbulence the rate of star-forming collisions also increases and stars of decreasing mass are produced in increasing numbers.

Thus far the theoretical model agrees with the Williams and Cremin results, except for explaining the formation of more massive

[a]Woolfson, M.M. (1979), *Phil. Trans. R. Soc. Lond.* **A291**, 219.

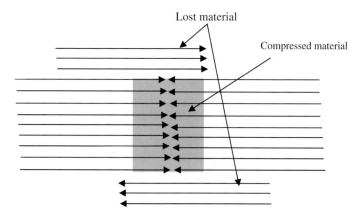

Figure 12.6 The head-on collision of two gas streams. The central material forms a compressed region with little rotation. The peripheral material is not retained.

stars. A star produced by the head-on collision of two streams of material will tend to have a small amount of rotation, the reason for which is illustrated in Figure 12.6. The colliding regions form the star and whatever rotation it has will be due to variation of speeds within the streams' cross sections. The peripheral material, which corresponds to a spin in a clockwise direction, moves quickly with respect to the compressed region and so is lost. The estimated spin rate of the Sun given by this mode of formation is a few times the present observed rate but there are mechanisms that can reduce the initial rotation rates to a limited, but sufficient, extent. One mechanism involves the presence of a stellar magnetic field together with emission by the star of large numbers of charged ions, mostly protons. This slow-down mechanism depends on a physical quantity called angular momentum but here we relate the slow-down mechanism to a simple experiment which can be done with a rotating chair. Imagine sitting in such a chair with two heavy weights, one in each hand. Then the chair is set spinning (if you try this, be careful!). Now stretch out your arms so the weights are further from the spin axis of the chair. You will find that the chair spins more slowly. Bring the weights in again and the spin is faster. However, if you drop the weights with your arms extended then the chair continues to spin more slowly. Let us see how to relate this to the star.

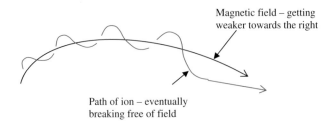

Magnetic field – getting
weaker towards the right

Path of ion – eventually
breaking free of field

Figure 12.7   The path of an ion in a varying magnetic field.

A magnetic field has strength and a direction at any point. The strength is a measure of the magnitude of the force it exerts on magnetic material or a moving charged particle and its direction is shown by a compass needle placed in the field. If an electrically-charged particle moves in a magnetic field, its path is a spiral around the direction of the field. The stronger the field, the tighter the spiral is (Figure 12.7). The field weakens with increasing distance from the star so the radius of the spiral increases until, when the field is sufficiently weak, the ion breaks away from the field altogether. The magnetic field of a star rotates with the star and the ions are carrying mass outwards, rather like the weights being extended in the rotating-chair experiment. When the ions break away it is similar to dropping the weights and the star, like the chair, is left rotating more slowly than it was previously. Unlike for the rotating chair, where a single pair of weights was dropped, for the star the discarding of mass takes place continuously, but the principle still applies. Early in the life of stars the rate of loss of charged ions and the strength of the magnetic field is much greater than they are when the star reaches the main sequence. By this process a slow-down in the rotation of the Sun from up to ten times its present value is possible.

Now we consider the formation of stars with more than 1.35 times the solar mass. This is explained by the accretion of cloud material by the newly-formed star, something that takes place more readily in the central denser regions of the cloud. Another mechanism for giving higher mass stars is for newly-formed protostars to collide and coalesce with a previously-formed protostar. In both cases the effect of adding material to the existing star increases the rate of spin.

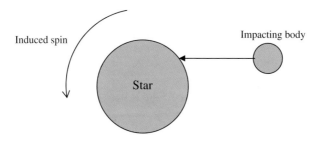

Figure 12.8   Spin imparted by an impacting body.

Figure 12.8 shows the effect of material impinging onto an already formed star — it does not matter whether the added mass is a compact body, such as a smaller protostar, or a stream of gas. The effect is that the sideswipe addition acts to spin the star, in an anticlockwise direction. Of course, if several additions of material are made then their spin contributions will be randomly related to each other and will tend to cancel each other. However, it can be shown theoretically that, the more massive the star becomes, the greater will be its spin rate. The 1979 calculations relating to the formation of more massive stars gave excellent agreement with the observational results shown in Figure 12.4.

For forming very massive stars, with mass greater than about ten times the solar mass, another consideration arises, something examined by the astronomers I.A. Bonnell, M.R. Bate and H. Zinnecker; for these stars, it is *essential* that they should be formed by the coalescence of smaller compact bodies. The model we have for forming a star involves, in the final stages, a compressed, but relatively cool, mass of gas starting its collapse towards the main sequence. In reality collapse is not uniform; the central regions fall in more quickly and outer material then joins the central condensation. However, for a very massive star the temperature generated by the collapsing core becomes so high that outer material in a diffuse form, rather than joining the core, is driven outwards by the pressure of the intense radiation. In that case, even if a high density region of the required mass could form, a star of very large mass could not form by its collapse. Indeed, it is a matter of observation that massive stars form only in the denser regions of a star-forming cloud where collisions will be more common. We shall have more to say about that in the next chapter.

12.6 The Formation of Binary Systems

Finally we consider why so many binary systems should form. In most binary systems, the two stars are so close that it is not possible to resolve them separately — through a telescope they look like a single body. However, the Doppler Effect can be used to recognize the existence of two bodies. In Figure 12.9 we show a simple arrangement that demonstrates the way that this is done. The plane of the stars' orbits is taken to contain the Earth and the motions of the stars at a particular time are shown. If the centre of the orbit were stationary with respect to the Earth then there would be a blue shift in light coming from A and a red shift in that coming from B. Hence a particular spectral line would appear as two separate lines, one on each side of the normal position. Even if the plane of the orbit does not contain the Earth and the centre of the stars' orbits is not stationary with respect to the Earth, the splitting of spectral lines still reveals the presence of the two stars and can be used to determine the characteristics of the binary system. Binary systems that can only be recognised and analysed in this way are known as *spectroscopic binaries.*

A close spectroscopic binary can come about if, by the process described in Figure 12.8, a star achieves a high rate of spin while it is collapsing. This mechanism was described by James Jeans in 1916. The stages in the evolution of the star, as it collapses, are shown in Figure 12.10. Initially, the star takes the form of an *oblate spheroid,*

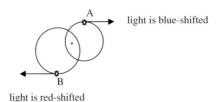

Figure 12.9   Observations of a spectroscopic binary.

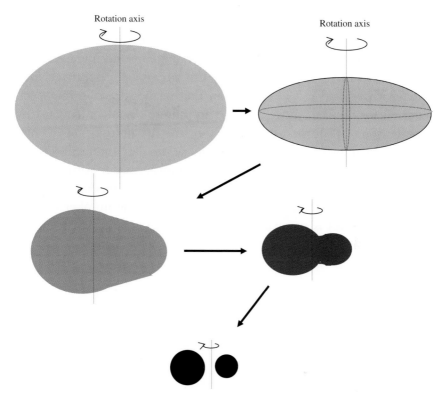

Figure 12.10   Stages in the development of a spinning, collapsing star to form a close binary system.

rather like a squashed football. Next it goes into the form of an *ellipsoid*, like a misshapen rugby ball, followed by a pear shape with a pointed and a blunt end. Then a neck appears towards the pointed end and, finally, the star undergoes fission into two separate smaller stars orbiting closely around each other — forming a close, spectroscopic binary pair.

There is a mechanism, based on tidal effects, which could increase the distance between two close binary stars. However, it cannot explain the distances between some binary stars, so far apart that they are easily resolved. In *visual binaries*, which are much rarer than spectroscopic binaries, the distances between the stars can be thousands of astronomical units. The most probable mode of formation of such systems, and also systems containing more than two well-separated stars, is by gravitational capture in an environment where stars are comparatively close together. Where stars mill around under gravitational forces in close proximity they can occasionally approach each other with a relative speed less than the escape speed — the speed required for them to move apart to an infinite distance. In that case the two stars will form a binary pair. In the dynamically-active environment that produced the visual binary, further interactions may either break it up again or, less likely, add another star to form a three-star system. This can only happen with reasonable probability in an environment where the stellar number density (the number of stars per unit volume) is comparatively high. How such an environment may form will be our next consideration.

Chapter 13

A Crowded Environment

13.1 Embedded Clusters

A dark, cool cloud within which a galactic cluster of stars is forming would be in a state of collapse. Superimposed on the chaotic turbulence there is a general inward flow both of the material of the cloud and of previously formed stars in all stages of development. As the cloud collapses so turbulence increases, leading to an accelerating rate of star formation. With the passage of time, the central region of the cloud becomes occupied by an ever-increasing number of protostars and stars in ever-closer proximity. This state of affairs is actually observed in star-forming regions. Figure 13.1 shows the Trapezium Cluster, situated in the Orion Nebula, which is an active site of star formation.

In the core of the Trapezium cluster, where stars are still being formed, the density of stars is very high, of order 1,000 stars per cubic light year. By contrast, in an open cluster like the Pleiades (Figure 1.3) there is about one star per cubic light year, and the density of stars in the vicinity of the Sun, a field star, is about 1 star per 300 cubic light years. Dense stellar environments have now been observed in many star-forming regions, and developing clusters in this state are known as *embedded clusters*. It is inferred that the embedded state is a forerunner of all normal open clusters.

Stars in the embedded state of a forming cluster are 'embedded' in a great deal of gas — hence the terminology. The gravitational effect of this gas, which accounts for more than one half of all the

Figure 13.1   The Trapezium Cluster, situated in the Orion Nebula, showing four very bright stars (Hubble).

mass in the cluster, binds the stars together and promotes their collapse. In the evolving cluster, there will be a few stars with masses more than ten solar masses, and these will reach the supernova stage in a few million years (Table 11.1). The blast effect of a supernova explosion expels much of the gas in its vicinity out of the cluster. The combined effect of several supernovae is to reduce the gravitational pull of the gas and the embedded cluster then begins to expand. The average duration of the embedded stage of a cluster from formation to the beginning of expansion, is about five million years — about the lifetime of a ten solar-mass star.

It has been estimated that only 5 per cent or so of expanding clusters will give the formation of an open cluster like the Pleiades. The remaining 95 per cent will continue to expand until all the constituent stars have become field stars. Hence it is almost certain that the Sun was a member of an embedded cluster, all the stars of which dispersed to become field stars. We say *almost* certain because there is another possibility. When a cluster forms — either open or globular — the stars are milling around within the cluster under their mutual gravitational forces. They gain and lose energy in this process and, occasionally, a star near the boundary of the cluster will move outwards with sufficient speed to escape and so become a field star. We refer to this process as the *evaporation* of the cluster, an apt description because it diminishes the cluster in much the same way that liquid evaporation diminishes a pool of water. For water, the individual molecules have a wide range of speeds and any molecules near the surface with more than some minimum speed are able to overcome the attraction of the liquid surface and so escape. Eventually a stellar cluster will be left as a stable system of two or more stars. All open clusters eventually disperse in this way since their average lifetime is a few hundred million years, although with a large range of lifetimes. A small cluster may last a few tens of millions of years but a large cluster could last a billion years or more. For globular clusters, containing hundreds of thousands of stars, the theoretical lifetimes are in the range 10^{12} to 10^{14} years, much greater than the age of the universe. We can be confident that all the globular clusters that have ever been formed are still around today.

13.2 Interactions Between Stars

In the embedded state, the number density of stars is so high that interactions between them are not uncommon. As an example, with the Sun in its present environment, and with a relative speed of about 30 kilometres per second for stars in the Sun's vicinity, the time between approaches of other stars to within one tenth of a light year (~6,000 astronomical units) from the Sun is about one hundred million years. Within a dense embedded cluster, where stars are much closer together but have lower relative speeds, the time for a particular star to be approached by another star to within that distance is one thousand years. Hence, in the average lifetime of an embedded cluster, five million years, there would be 5,000 such interactions for each star. Of those 5,000 interactions, on average one will be within 350 astronomical units, about 12 times the distance of the Sun from Neptune but a third, or less, of the radius of a newly-formed protostar. We can infer that an embedded cluster containing a few hundred stars will be a bustling environment with significant stellar interactions continuously going on. By the time the embedded cluster eventually expands or disperses there should have been some significant outcomes from all that activity.

The problem of forming very massive stars, with mass greater than about ten solar masses, was referred to in Section 12.5. Such stars cannot be produced by the collapse of a single cloud of material with the required mass because the central core would be so luminous that it would drive out the outer material. For this reason Burnell and his colleagues proposed that massive stars can only be produced by the accretion of low-mass stars, or protostars, onto a previously formed star. Observations indicate that the most massive stars are formed in the inner cores of embedded clusters. The Jeans critical mass in such a dense region would be small, perhaps one-third of a solar mass, so the idea that massive stars could form there by direct collapse of a forming protostar is not really tenable, even without the problem of radiation from the core. However, a star of, say, one solar mass, formed earlier, that happened to wander into the inner core region while that region was a prolific source of less massive star formation would be able to accrete many of the smaller bodies. The larger its mass the more it would attract other bodies to join it, and the more bodies that joined it the larger would be its mass. While the radiation from the growing massive star would prevent *diffuse* gaseous material impinging onto it, the radiation would have little effect on the arrival of *compact* bodies. This is a self-consistent picture for the formation of massive stars. Only in the cores of embedded clusters could sufficient accretion occur to build up a massive star and massive stars are almost always found in the central regions of dense stellar clusters.

Another consequence of the embedded state of a cluster is found in the frequency of binary systems. The formation of spectroscopic and visual binary systems is described in Section 12.6. Close spectroscopic binaries are by far the most common and are also the most resistant to being disrupted. A German astronomer, Pavel Kroupa (b. 1963), has investigated the idea that the greater the density of stars in an embedded cluster, and the longer a binary system stays in that environment, the greater the probability that the binary system will be disrupted by tidal effects due to other stars. It is believed that the proportion of stars produced as binary systems, when stars are first formed, is greater than the proportion now observed as field binaries or in dense clusters and that some of the original binaries were disrupted while the cluster was in the embedded state.

The idea that interactions between stars are common in embedded clusters is now well established. Later, another kind of interaction will be described — one that can produce planets. However, before describing this we first consider the structure of the Solar System and recent observations of planets around other stars.

The Solar System

Chapter 14

Understanding the Nature of the Solar System

14.1 Ptolemy's Earth-Centred Model

From antiquity, mankind had observed the sky and wondered about the significance of the myriad points of light — the stars. As night progressed, so the stars rotated in the sky, maintaining their relative positions so that prominent patterns, the constellations, could always be recognised. Once men took to the seas, or travelled on featureless plains, stars were used as a means of navigating. In the northern hemisphere, two of the seven stars of *The Big Dipper* (Figure 14.1), part of the constellation *Ursa Major* (Big Bear), pointed to *Polaris*, the pole star, which gives the direction of true north.

There were exceptions to this fixed pattern in the sky. Five 'stars' wandered amongst the background of fixed stars; they became known as planets, from the Greek word *planetos* (πλανετοσ) meaning *wanderer*. Eventually it was realized that these bodies, together with the Sun, Earth and Moon, formed some kind of coherent system. Since stars apparently circle the Earth, and there is no sensation of motion of the Earth, the idea became established that the Earth was stationary and that the stars, Sun, Moon and planets all moved around it. The Alexandrian Greek astronomer Ptolemy (Figure 14.2) devised a scheme for explaining the motions of the planets as seen from the Earth, with Earth as the central body.

Figure 14.1   The Big Dipper (also known as The Plough). The arrow, following the direction of the two right-hand stars, points towards Polaris, the next bright star along that line. (Note: the stars have been artificially enhanced to make them easier to see.)

Figure 14.2   Ptolemy (100–170 AD).

As seen from Earth, the rotations of the stars, Sun and the Moon are at a uniform rate. These motions can be represented by travel in a circle at a constant speed around the Earth, something easy to visualise and to understand. The motions of the planets were much more difficult to understand — instead of moving smoothly in one direction against the background of the stars they occasionally made looping motions (Figure 14.3).

In Greek philosophy the concept of perfection was important, and the shapes that had this quality of perfection were the circle and

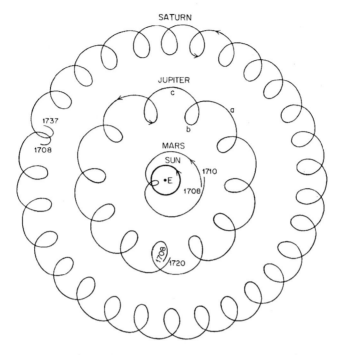

Figure 14.3 Motions of Mars, Jupiter and Saturn as seen from Earth.

the sphere. They believed that nature embraced this concept so that, without supporting evidence, they made the correct assumption that major heavenly bodies were spheres. For the same reason, it disturbed them to find that, alone of all the bodies in the firmament, the planets were not moving in circular paths about the Earth. Ptolemy's description of planetary motions dealt with this disquiet by describing the motions of the planets as a superposition of two separate circular motions. This is illustrated in Figure 14.4. A point called the *deferent* circles the Earth at a constant speed, while the planet moves at a constant speed on a circular path, called the *epicycle*, around the deferent. These combined motions explained the positions of the planets as seen from Earth to the accuracy with which the positions were then measured. The motion was complicated, but the laws that governed planetary motion were not known and this description had the virtue of explaining what was observed.

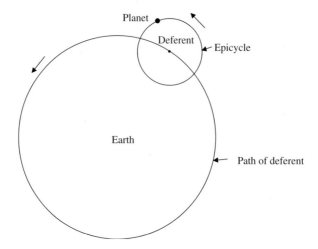

Figure 14.4 Ptolemy's description of planetary motion.

This Earth-centred description of the Solar System was accepted for the next 1,400 years, especially as it agreed with the description of creation as described in the biblical book of Genesis. First the heavens and the Earth were created, later the Sun and the Moon and finally, as the pinnacle of the creation process, man. The Earth and mankind at the centre of the Universe seemed quite natural to believers in the Jewish-Christian-Muslim tradition. Actually, a Greek philosopher, Aristarchos of Samos (310–230 BC), had suggested that the Sun and the stars were fixed in position and that the Earth moved round the Sun, but this idea seemed so preposterous that it was rejected — at least for the next 1,800 years.

14.2 The Copernicus Heliocentric Model

The seminal step in understanding the true nature of motions in the Solar System was due to a Polish cleric, Nicolaus Copernicus (Figure 14.5), whose education covered the fields of medicine, mathematics and astronomy. Although he spent a brief period in Bologna and Rome, his main service to the church was in Warmia, near the Baltic coast and 150 kilometres from his birth city, Torun. Copernicus

Figure 14.5   Nicolaus Copernicus (1473–1543).

was greatly influenced by the writings of Ptolemy who had written 13 books, called the *Almagest* (the Latinised form of an Arabic name meaning *The Great Book*), which contained all the astronomical knowledge of the 2nd century AD. Copernicus made improved observations of planetary motions, on the basis of which he decided that Ptolemy's description in terms of geocentric (Earth-centred) motion was not only complicated but also disagreed with his new observations. He constructed a heliocentric (Sun-centred) model in which all planets circled the Sun. Like Ptolemy he believed that orbits should be based on circles, but his observations showed that the angular speeds of a planet's motion around the Sun varied slightly. He solved this problem by having the centre of the circular motion displaced from the Sun. If the planet went round the circular orbit at a constant speed then when it was closest to the Sun its angular speed would be higher than average and when furthest from the Sun the angular speed would be lower than average. Even using this scheme there were still some residual discrepancies with observations so he introduced epicycles, although they were tiny compared with those assumed by Ptolemy. Copernicus also rejected the idea that stars were all in rotation around the Earth, or even the Sun, and stated that it was Earth's spin around its polar axis that gives this impression.

Copernicus described his model in a book *De Revolutionibus Orbium Coelestium* (On the Orbits of Heavenly Spheres), which was published shortly before he died. He dedicated it to Pope Paul III but the church took little notice of what he had written, something that, as later events were to show, was just as well for Copernicus.

14.3 Tycho Brahe

The next notable contributor to an understanding of the nature of the Solar System was a Danish nobleman and astronomer Tycho Brahe (1546–1601). He had connections to Danish royalty, and was a haughty, quick-tempered and quarrelsome man. As a student his nose was sliced off in a duel and thereafter he wore a false nose made of gold and silver. When young he found favour with the king, Frederick II, who presented him with the island of Hven, between Denmark and Sweden, and the resources to build an observatory. Telescopes had not been invented and the observatory, Uraniborg (roughly translating in Danish as *sky castle*), was equipped with line-of-sight instruments that gave very precise directions of astronomical bodies. Figure 14.6 shows a huge brass quadrant mounted against a wall. On the wall is a picture of Tycho and the man himself is just visible at the right-hand edge making an observation. Two other men are seen, one recording time from a clock and another, seated at a table, noting the observations.

Tycho did not accept the Copernican model of the Solar System; his model combined elements of the ideas put forward by both Ptolemy and Copernicus. In this hybrid model all the planets, except the Earth, went round the Sun but the Sun and the Moon orbited the Earth. This model correctly explains the relative motion of solar-system bodies and, given that no theory then existed to explain how the bodies moved, that was all that was needed at the time.

True to character, Tycho ran into trouble when his patron, Frederick II, died in 1588. Tycho mistreated the inhabitants of Hven, who were his tenants, and Frederick's successor, Christian, resented Tycho's arrogance, bad behaviour and ingratitude for the generosity that had been shown to him. In 1599, under pressure, Tycho left Hven to become the Imperial Mathematician at the court of Rudolph II in Prague. There he compiled accurate tables of planetary motions, assisted by a very talented young man, Johannes Kepler (Figure 14.7).

Figure 14.6   Tycho Brahe's quadrant instrument.

Figure 14.7   Johannes Kepler (1571–1630).

14.4 Johannes Kepler

When Tycho died Kepler inherited all his observational data and initiated a great project using this material to determine the exact motions
of the planets. Kepler concentrated his efforts on Mars, which was easy
to observe and whose orbit departed appreciably from circular motion.
After eight years Kepler formulated two laws of planetary motion and
after a further nine years he discovered the third law. These are:

(i) Planets move in elliptical orbits with the Sun at one focus.
(ii) The radius vector sweeps out equal areas in equal times.
(iii) The square of the period is proportional to the cube of the mean
distance of the planet from the Sun.

An ellipse is an oval shape, several of which are shown in Figure 14.8.
Within an ellipse there are situated two special points, the *foci*, shown

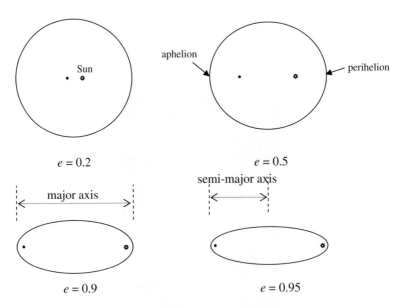

Figure 14.8   Elliptical orbits with various eccentricities. The two foci are shown
with one made more prominent to indicate the position of the Sun if
the ellipse was a planetary orbit. Also shown are the major axis and the
semi-major axis.

in the figure. Ellipses vary from near circular to very elongated forms and the degree of departure from a circle is given by a quantity called the *eccentricity* (e) of the ellipse. A circle is a special case of an ellipse with $e = 0$. As e increases towards a value of 1, the ellipse becomes more and more elongated. For $e = 1$ the ellipse becomes another geometrical shape called a *parabola*. Most planetary orbits are close to circular but Mars, Kepler's planet of interest, has an eccentricity of 0.095. The closest orbital point to the Sun is known as the *perihelion* and the furthest point the *aphelion* (pronounced 'afelion'). For the eccentricity of Mars the ratio of the furthest distance to the Sun, to the closest distance to the Sun, is about 1.19. The eccentricity of an ellipse indicates its shape but not its size. This is given by the *major axis*, the long dimension of the ellipse, or, more commonly, the *semi-major axis*, one half of the major axis. For the Earth the orbital eccentricity is 0.0167 and the semi-major axis is 1.496×10^8 kilometres, which is the *astronomical unit*. What is surprising is the long time it took for Kepler to find that the form of the orbit was an ellipse. He was a skilled geometrician and the ellipse was a well-known geometrical shape.

Kepler's second law refers to the radius vector, the line joining the Sun to the planet. When the planet is further from the Sun it moves more slowly. Figure 14.9 shows the area swept out by the radius

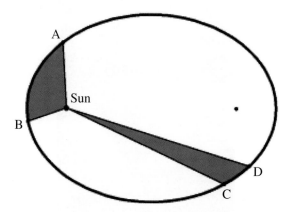

Figure 14.9 Kepler's second law. The planet takes the same time to go from A to B as it takes to go from C to D. The shaded areas shown are the same.

vector in two equal periods of time and the law tells us that these areas are equal.

For the third law, the period is the time taken for the planet to complete one orbit and the square of this time is proportional to the cube of the semi-major axis.

14.5 Galileo Galilei

A famous contemporary of Kepler was Galileo Galilei (Figure 14.10). He was a mathematics professor in Pisa at the age of 25 and also taught at the University of Padua. His major interests were mechanics in general and planetary motion in particular. Galileo and Kepler corresponded and agreed that they both favoured the Copernican heliocentric model.

In 1600 an event occurred that was to have a dramatic effect on Galileo's life. A Dominican monk and philosopher, Giordano Bruno, proposed that stars were like the Sun and had accompanying planets that were inhabited by other races of men. This challenged the doctrine of the Church concerning the unique status of mankind, created in God's image. Bruno was brought before the Inquisition and was ordered to recant, but he refused to do so. He was burnt at the

Figure 14.10   Galileo Galilei (1562–1642).

stake for his heresy and, some time thereafter, there began a rooting out of astronomical literature that might be seen to be heretical. *De Revolutionibus* and publications by Kepler were added to the *Index Librorum Prohibitorum*, the list of books that it was prohibited for Catholics to read. It became dangerous to be seen as a supporter of the heliocentric theory, which presented Galileo with severe problems. The Lutheran Church, which dominated the north of Europe, was equally opposed to the Copernicus model but, since it was nothing like as powerful in controlling domestic affairs as was the Catholic Church, Kepler was able to continue his work without undue hindrance.

At the beginning of the 17th century, Hans Lippershey, a Dutch spectacle maker, designed the first telescope. In 1608, Galileo made a telescope, with the encouragement of the Venetian Senate, which saw that it could have commercial and military applications. Galileo soon turned his telescope towards the sky and he made many important observations. He saw mountains on the Moon, the rings of Saturn (although he did not recognize their nature) and found four large satellites of the planet Jupiter, now called the Galilean satellites in his honour. Galileo saw these satellites as a smaller version of the Copernican model, which reinforced his belief in the heliocentric theory — although that was not a proof that the Copernican model was correct. However, one observation that Galileo made *did* support the Copernican model and also showed that the Ptolemy model was untenable. This involved the planet Venus, which, from the Earth, is always seen not far from the Sun.

In the Ptolemy model the Sun goes round the Earth in a circular orbit with a very small epicycle; if we take it as being of zero radius then the Sun would be coincident with the deferent (Figure 14.4). The only way that Venus is always seen close to the Sun is if its deferent was on the Earth–Sun line and then, as Venus went round its epicycle it would be seen going from the left to right of the Sun, and back again, as is actually observed. This is shown in Figure 14.11, which shows the view from the Earth towards the Sun and the path of Venus.

From Figure 14.11 it is clear that Venus, as viewed from the Earth, can never be seen as a 'full' Venus, i.e. with a face completely

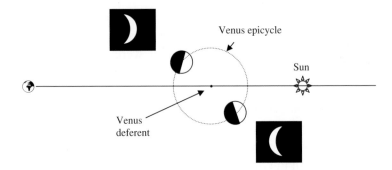

Figure 14.11   Views of Venus as seen for the Ptolemy model.

illuminated by the Sun. It will always be seen in a crescent phase. Figure 14.12 shows the situation that occurs with the Copernican model. In this case Venus never strays too far from the Sun's direction because it is on an inner orbit. However, when Venus is in position A the rear of the planet is illuminated and the planet is effectively invisible. A move slightly to the left or right will give a crescent phase as is seen in the rectangular frame. Conversely, if Venus and the Earth are on the opposite sides of the Sun (position B) then a fully illuminated face of Venus is seen — a 'full' Venus. In this case, since Venus is further from the Earth than it was at A, the size of the full Venus is much smaller than the size of the crescent Venus.

Galileo's observations showed that the Copernican theory was the correct one. He saw Venus either large in a crescent phase or small in a full phase and that was completely inconsistent with the geocentric theory. What could Galileo do? On the one hand he was a devout man (despite fathering many illegitimate children) who did not wish to flout the Church's rule, but he was no Bruno and did not wish to die, or to suffer in any way, for his scientific beliefs. On the other hand his scientific observations were unambiguous in what they indicated — Copernicus was right and Ptolemy was wrong.

Galileo decided on a subterfuge. In 1632 he published a book, *Dialogue on Two World Systems,* in which two characters discussed, in a supposedly dispassionate way, the respective merits of the geocentric and heliocentric models. Salviati presents the case for the

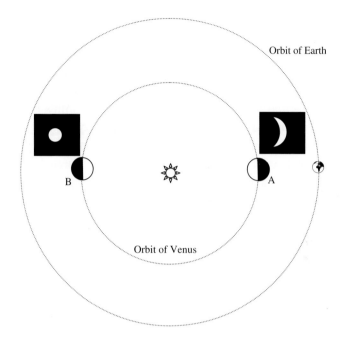

Figure 14.12 Phases of Venus as seen from the Copernican theory.

Copernican model as Galileo himself would have presented it. Simplicio argues for the Ptolemy model, but there is no doubt to any reader that he is intellectually outgunned by Salviati. A third character in the book, Sagredo, is an intelligent layman to whom they address their arguments and who occasionally asks questions. The book, with its convoluted arguments, is hardly comprehensible to a modern reader, but the reader of the time would have had no doubt which case was the stronger. The Inquisition was not fooled and within a few months Galileo was forced to appear before them and to recant his heretical views. Galileo was never subjected to harsh treatment — he was sensible and gave ground to threats of violence — but he was under virtual house arrest for the remainder of his life. Truth had given way to dogma, prejudice and brute force in this instance. Later there were to be many conflicts between science and religion but they were carried out by force of argument rather than by threats of force.

14.6 Isaac Newton

To complete the topic of understanding the nature of the Solar System, we now come to, arguably, the greatest scientist of all time, Isaac Newton (Figure 5.1). Newton's contributions ranged over many fields — mathematics, optics and hydrodynamics to name just some. Our interest is his discovery of the force that governs the motions of the planets and other bodies, the force of gravity. The inverse-square law of gravitation states that the force between two bodies is proportional to the product of their masses and inversely proportional to the square of the distance between them. This force explains the motion of the Moon round the Earth and the motions of the planets round the Sun. It could also explain the fall of apples from a tree, the observation of which is claimed to have stimulated Newton's development of the theory of gravitation. Mathematical analysis shows that Kepler's three laws of planetary motion follow from the inverse-square law of gravitation. Now, scientists knew *how* bodies moved in the Solar System and also knew *why* they moved in that way. There were many discoveries about the Solar System still to be made, but thenceforth the basic mechanics of the Solar System was completely understood.

Chapter 15

Introducing the Planets

15.1 An Overall Description of the Planetary System

There are planets around stars other than the Sun, but only in the Solar System can we study planets and other constituent bodies in some detail. To systematise our description of the Solar System we start with the largest bodies and work down the size scale. This cannot be done perfectly as there is size overlap of some types of body — for example, some satellites are larger than the smallest planet — but in general the size sequence will be followed. In this chapter we deal with the planets.

The Sun, a typical main-sequence star, is the dominant member of the system, accounting for 99.86% of its total mass. Next in the size sequence are planets, illustrated in terms of their relative sizes in Figure 15.1. Planets are divided into two groups, the large major planets, Jupiter, Saturn, Uranus and Neptune, and the much smaller terrestrial planets Mercury, Venus, Earth and Mars. These two groups are distinguished not only by their sizes and masses but also by their compositions. The terrestrial planets, like the Earth, are primarily spheres of silicate rock with iron interiors, and their atmospheres, which all except Mercury possess, are minor components. By contrast, the major planets are mainly of gaseous composition, although probably with silicate-iron cores. Table 15.1 gives the masses, radii and mean densities of the planets.

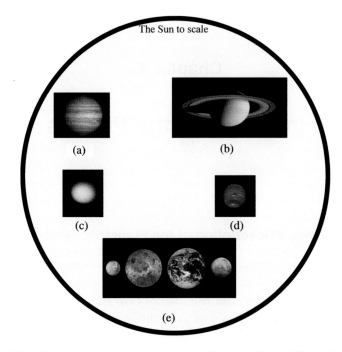

Figure 15.1   Solar-system planets giving an indication of size. The major planets are: (a) Jupiter, (b) Saturn, (c) Uranus and (d) Neptune. (e) The four terrestrial planets are, from left to right, Mercury, Venus, Earth and Mars. They are scaled up in size by a factor of 10 relative to the major planets.

Table 15.1   The physical characteristics of the planets.

Planet	Mass (Earth units)	Diameter (kilometres)	Density (kilograms per cubic metre)
Mercury	0.0533	4,879	5,427
Venus	0.8150	12,104	5,243
Earth	1.0000	12,756	5,515
Mars	0.1074	6,794	3,933
Jupiter	317.8	142,984	1,326
Saturn	95.16	120,536	687
Uranus	14.5	51,118	1,270
Neptune	17.2	48,400	1,638

Table 15.2 The orbital characteristics of the planets.

Planet	Semi-major axis (astronomical units)	Eccentricity	Inclination (°)	Orbital period (years)
Mercury	0.387	0.2056	7.0	0.2409
Venus	0.723	0.0068	3.4	0.6152
Earth	1.000	0.017	0.0	1.0000
Mars	1.524	0.093	1.8	1.8809
Jupiter	5.203	0.048	1.3	11.8623
Saturn	9.539	0.056	2.5	29.458
Uranus	19.19	0.047	0.8	84.01
Neptune	30.07	0.0086	1.8	164.79

The two groups of planet are also distinguished by their locations (Table 15.2), the terrestrial planets being closely spaced within the inner part of the system while the major planets are much further out and well separated.

The Earth's orbital plane, *the ecliptic*, is the reference plane for defining the orbital planes of other planets. These are given by their *inclinations*, the angles that the planets' orbital planes make with the ecliptic. The smallest planet, Mercury, has the most extreme orbit, with both the highest eccentricity and the highest inclination. The reason that Kepler did not use Mercury's orbit for his analysis of planetary motion is that, being so close to the Sun, it is difficult to observe and its orbit was not accurately known at that time.

Another important feature of planets is their spin periods. In the case of the Earth this is 23 hours 56 minutes that, in conjunction with the motion of the Earth around the Sun, gives a 24-hour day. The orientation of the spin axes is yet another important characteristic. If the spin axis of a planet were perpendicular to its orbital plane, and the orbit was circular, or nearly so, then there would be no seasonal effects on the planet; a particular region of the planet would have the same daily pattern of exposure to the Sun at all points of its orbit at all times. The tilt of the spin axis is given by the angle it makes with the normal (perpendicular) to the orbital plane and, with this definition, the Earth's spin axis is tilted by 23½° — giving seasonal effects as shown in Figure 15.2. In the northern summer the northern hemisphere is tilted towards the Sun and so gets greater exposure to solar radiation. Conversely in the northern winter the northern hemisphere is tilted away from the Sun and gets less exposure. The times when the spin axis is in the vertical plane containing the Sun are the *solstices* — in the northern hemisphere the winter solstice is around December 21st and the summer solstice around June 21st.

The spin periods and spin-axis tilts of the planets are given in Table 15.3. For Venus and Uranus the axial tilt is greater than 90°, so that, in projection on the plane of the orbit, they are spinning in the

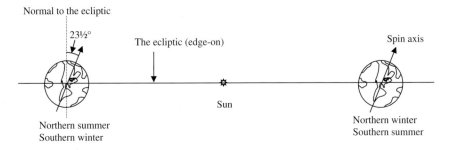

Figure 15.2   The Earth at positions six months apart, at the solstices.

Table 15.3   The spin periods and axial tilts of the planets.

Planet	Spin period	Axial tilt (°)
Mercury	58.6 days	0.01
Venus	243 days	177.4
Earth	23.9 hours	23.5
Mars	24.6 hours	25.2
Jupiter	9.9 hours	3.1
Saturn	10.7 hours	26.7
Uranus	17.2 hours	97.8
Neptune	16.1 hours	28.3

opposite sense to the other planets. Most rotations in the Solar System are anticlockwise when looking down on the plane of the system from the north. Such rotations are termed *direct*, or *prograde*, while rotations in the opposite direction are *retrograde*. The spins of Venus and Uranus are retrograde.

15.2 The Major Planets

Jupiter is the largest planet, with mass two-and-a-half times that of all other planetary masses combined and 318 times that of the Earth — but only about one-thousandth that of the Sun. It is a gaseous sphere, mainly of hydrogen and helium but with minor components of molecular species made up of various combinations of hydrogen, carbon, nitrogen and sulphur. Its silicate-iron core is thought to have a mass somewhere between five and ten times that of the Earth. Surface features consist of oval swirls of material, varying in colour, that create the bands seen in Figure 15.1(a). The most prominent feature is the *Great Red Spot*, a large oval feature several times bigger than the Earth, which is at the lower right-hand side of the image. This is a storm that has been seen on Jupiter for more than 350 years. The smaller swirls are also storms but these come and go on a timescale of a year or so. Jupiter has the fastest spin rate of any planet with a spin period of 9 hours 55 minutes, varying slightly with latitude.

Saturn is a smaller version of Jupiter, with a mass 95 times that of the Earth. Its surface features are similar, although with nothing to show as spectacular and long-lasting as the Great Red Spot. Saturn's rate of spin is second only to that of Jupiter, with a period varying from 10 hours 15 minutes at the equator to 10 hours 38 minutes closer to the poles. A striking feature of the planet is its low density, one-half that of Jupiter and 70% that of water. Its material is highly concentrated towards the centre, with outer material being quite diffuse. This distribution of material, combined with its rapid spin, leads to discernable flattening of the planet along its spin axis. The most spectacular feature of this planet is its ring system (Figure 15.1(b)); it is incredibly thin, a kilometre or so, consisting of numerous small bodies, with size from grains up to a few metres, which are of silicate

Figure 15.3   A detailed view of the rings of Saturn (SSI, JPL, ESA, NASA).

or ice composition. The pictures of the rings taken by the Voyager I spacecraft in 1980 revealed their very detailed structure (Figure 15.3).

Galileo observed Saturn's rings and referred to them as 'ears' attached to the planet; his small telescope could not show their true nature. He was amazed when, two years later, the 'ears' disappeared. As Saturn and the Earth orbit the Sun, from time to time the rings are edge-on to the Earth, and hence apparently disappear. All the major planets have ring systems but, with the exception of Saturn, so sparse in nature that they cannot be observed from Earth. It is thought that Saturn's rings were caused by a body, perhaps a satellite, drifting in towards the planet and then being torn apart by strong tidal forces. We shall refer later to tidal forces and how they operate.

Uranus has a radius just over one-third that of Jupiter and mass 14.5 times that of the Earth. It contains considerable amounts of hydrogen and helium and its colour, greenish blue, is due to the presence of methane, CH_4, in its atmosphere. Uranus has a bland, featureless appearance that made the measurement of its spin rate difficult to determine until it was visited by spacecraft. The period of spin varies with latitude, being 17 hours at the equator down to 15 hours at higher latitudes. This higher rate of spin at higher latitudes is the opposite of what is observed for Jupiter and Saturn. The most extraordinary thing about Uranus is its axial tilt, inclined at 98°, which makes the spin axis only 8° from the orbital plane. This curious arrangement must be related to some event in its history.

The final major planet, Neptune, is sometimes described as the twin of Uranus. Its mass, 17.1 Earth masses, is greater than that of Uranus while its radius is less, so that it has a significantly greater density. The tilt of its spin axis, 28.3°, is similar to that of the Earth, Mars and Saturn — large, but not unusually large. Its surface has one notable feature, the *Great Dark Spot*, which can be seen at the centre of the image in Figure 15.1. This is a great storm system, similar to the Great Red Spot on Jupiter.

15.3 The Terrestrial Planets

The innermost terrestrial planet, Mercury, is close enough to the Sun to be affected by solar tidal forces. An interesting consequence of this is that its spin and orbital periods are related, with the spin period, 58.65 days, being exactly two thirds of the orbital period, 87.97 days. Consequently, there are two points on the surface at the equator, on opposite sides of the planet, which are directly under the Sun at alternate perihelion passages. Mercury is the only planet without an atmosphere and, since there are no atmospheric currents to transport heat, the temperature variation over Mercury is huge, from 90 K at the furthest point from the Sun to 740 K facing the Sun — a temperature high enough to melt lead, zinc and tin. A section of the surface of Mercury is shown in Figure 15.4. The left-hand side shows part of the boundary of the Caloris Basin, a large circular impact feature, some 1,500 kilometres in diameter, consisting of a series of ring structures. The surface of Mercury has been heavily bombarded, as is shown by the cratered surface, and between the craters are smooth plains.

Mercury has a very high density. Table 15.1 shows that it is only a little less dense than the Earth but, since the density of the Earth is affected by compression due to the pressure of its own mass, the intrinsic density of Mercury, with compression effects removed, is actually higher than that of the Earth. All terrestrial bodies are combinations of silicates and iron and the iron content of Mercury is proportionally higher than that of any other terrestrial planet.

Figure 15.4   The surface of Mercury with part of the Caloris Basin at the left-hand edge (NASA).

Venus, often regarded as a twin of the Earth, is actually very unlike the Earth in most characteristics. The carbon dioxide atmosphere is very thick and gives a surface pressure nearly 100 times that experienced on Earth. Because of the thick atmosphere, it is impossible to see the surface from outside by visible light but the surface has been extensively mapped by radar from spacecraft (Figure 15.5). There are mountains and flat areas, as on Earth, and three elevated regions that, if there were seas on Venus, would be continents. Actually, not only are there no seas on Venus, but it is a very arid planet. Since water is common in the Solar System it is almost inevitable that early Venus would have had a large complement of water. Venus is closer to the Sun than is the Earth and its early atmosphere would have contained a considerable amount of water vapour. Now, the Sun emits a great range of electromagnetic radiation and the effect of ultraviolet light is to dissociate (break up) water molecules,

Figure 15.5 A radar view of the surface of Venus.

H_2O, into OH + H. Hydrogen atoms are light so they move with speeds high enough for large numbers of them to escape from the planet's gravitational pull. Pairs of the OH radicals would then combine to give H_2O plus O, an oxygen atom, which would chemically react with any available material that was readily oxidised. In this way water would gradually be depleted until the present arid state was reached. The amount of water vapour in the atmosphere of Venus is about one thousandth of that in the Earth's atmosphere — so there is some water — but most has been lost.

This process explains another oddity about Venus — the large ratio of deuterium to hydrogen in the planet. Deuterium (Figure 7.2) is the isotope of hydrogen produced in the Big Bang. The ratio of deuterium to hydrogen (D/H) in the Universe at large is 2×10^{-5}, which is the ratio found in Jupiter, a massive planet that retained all the material from which it was formed. The ratio on Earth is much higher — D/H = 1.6×10^{-4} — and this indicates that in some way or other the hydrogen content of the Earth became enriched in deuterium. For Venus the ratio is one hundred times higher than on Earth, D/H = 0.016. This is explained as follows. The original water of Venus would have contained some partially deuterated water, HDO, which would have dissociated to give OH + D. Deuterium has twice the mass of hydrogen and would have moved more slowly, in fact too slowly to escape from Venus. Eventually the deuterium atom would have recombined with an OH to reform HDO. In this way hydrogen would be lost but deuterium would be retained, so steadily increasing the D/H ratio.

The temperature at the surface of Venus, 730 K, is similar to the maximum on Mercury, which is much closer to the Sun. The reason for this is the *greenhouse effect* due to the carbon dioxide atmosphere. The Sun's surface, at a temperature just below 6,000 K, emits most electromagnetic radiation in the visible to ultraviolet region. Carbon dioxide is quite transparent to these wavelengths of electromagnetic radiation so the radiation is transmitted through the atmosphere and heats up the surface. Unless all the absorbed radiation was radiated away from Venus then it would steadily heat up without limit. However, at the lower surface temperature of a few hundred degrees the radiation emitted has much longer wavelengths, in the infrared range, for which carbon dioxide is much less transparent. Consequently the surface heats up until the temperature is reached where the energy radiated equals the energy absorbed — and this corresponds to the high temperature at the surface.

We shall postpone our discussion of the Earth — we are familiar with its salient properties and, indeed, its future development under the impact of mankind's influence on it is much under discussion. We just note that if there were no greenhouse effect on Earth then the average temperature would be 255 K, below the freezing point of water (273 K). In such a circumstance life on Earth would be more difficult, although not impossible, but it is doubtful whether higher life forms, including man, would have developed. The greenhouse effect is clearly of benefit to our future existence but too much of it, going only a fraction of the way towards the Venus situation, would lead to our destruction.

The final terrestrial planet, Mars, has a reddish appearance, the colour of blood, and its name is that of the Roman god of war. The atmosphere is thin, with less than one per cent of the surface pressure on Earth, and like that of Venus is mostly carbon dioxide. The Martian surface can be seen through Earth-based telescopes that show icecaps that advance and recede with the seasons. At the end of the 19th Century, an Italian astronomer, Giovanni Schiaparelli (1835–1910), claimed to have seen 'canali' on the surface of Mars. The Italian word means 'channels' but was misinterpreted as 'canals' and so the myth grew of an advanced civilisation on Mars, exploiting water at the poles through a canal system. This led H.G. Wells to write his famous book *War of the Worlds* in 1898, which described the invasion of the Earth by Martians equipped with advanced war machines.

The view of Mars obtained from spacecraft and landers is less outlandish than that just described but is still incredibly interesting. The red colour of Mars can be crudely put down to rust since the surface is rich in the red oxide of iron, FeO (Fe is the chemical symbol for iron). The polar caps have a permanent component of water ice and a seasonal component of solid carbon dioxide, a substance known as *dry ice* that is manufactured on Earth for commercial use. Since Mars has a similar axial tilt to that of the Earth, it has seasons. In the northern summer carbon dioxide vaporises from the northern icecap and is deposited on the southern icecap and the reverse process happens in the southern summer (Figure 15.6). There is evidence that ice also exists below the surface away from the poles. The presence of

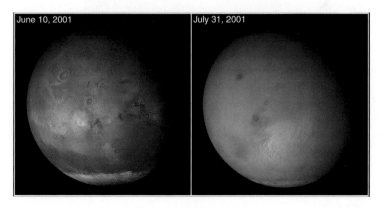

Figure 15.6. Views of a region of the Martian surface taken several weeks apart. The reduction of the icecap is seen. The earlier image shows a dust storm (lower-left centre).

large quantities of water, an essential ingredient for life, has promoted the idea that Mars may be suitable for colonisation — although there is still the problem of what to breathe!

There is evidence that, in the past, water on Mars was abundant, and in liquid form. Surface features resembling dried-up water channels (Figure 15.7), suggest a flow of water and a period when Mars had an appreciable atmosphere, giving a greenhouse effect that raised the temperature high enough to give a humid atmosphere, rain and liquid water. In August 2011 NASA announced that it had observed evidence of water flowing on Mars at the present time.[a] In the Martian summer, on sunlit slopes, dark tendrils, a few metres wide, are seen apparently coming out of the rocks and flowing a few hundred metres downhill. These tendrils disappear in the winter. It is postulated that it may be salty water with a low freezing point, coming from some warmer below-surface source. The presence of liquid water has renewed interest in the search for life on the planet.

Another extinct feature of Mars is volcanism. Many solar-system bodies, including the Earth, possess volcanoes, some active and some

Figure 15.7   This portion of the Martian surface shows several features that could be dried-up river beds (NASA).

---

[a] NASA, http://www.nasa.gov/mission_pages/MRO/news/mro20110804.html (Accessed 10/05/12).

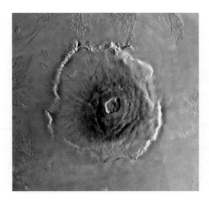

Figure 15.8 A view of Olympus Mons (Mariner 9, JPL, NASA).

extinct. Mars has many extinct volcanoes and the distinction of having the largest volcano in the Solar System, Olympus Mons (Figure 15.8). This towers 24 kilometres above the surrounding plain and is 600 kilometres across its base, dwarfing the largest volcano on Earth, Mauna Loa in Hawaii, which is 4 kilometres high and 120 kilometres across its base. Evidence of one-time extensive volcanism on Mars is apparent in the form of solidified lava flows running downhill from the volcanic centres.

The topology of Mars shows features that must relate to some aspect of its origin or history. The surface is divided into two distinct regions separated by a scarp (a steep slope) about 2 kilometres high, which runs at an angle of 35° to the equator. North of the scarp is a volcanic plain with few craters. Projectiles have bombarded all solar-system bodies over the lifetime of the Solar System; this area shows few craters because early damage was covered by flows of lava. It is estimated that these lava flows must have ceased about 3.8 billion years ago so that the few craters now visible have been produced since then. By contrast, the southern highland region is heavily cratered and shows the damage inflicted on it over the history of the planet. Figure 15.9 shows the topology; red indicates the highest regions and, passing through the spectrum, blue the lowest. The deep blue region in the south is the Hellas Basin, a depression 1,800 kilometres

Figure 15.9   The topology of Mars.

in diameter and 3 kilometres deep, caused by the impact of a huge projectile.

This concludes our brief survey of the planets. Much more could be said about each of them, but this broad-brush description suffices for us later to consider how planets, and all else in the Solar System, originated.

Chapter 16

Satellites Galore

16.1 The Satellites of Jupiter

When Galileo discovered the four satellites of Jupiter which now collectively bear his name, he saw them as a small version of the Copernican model of the Solar System. Both the Ptolemaic system and the Copernican system recognized the Moon as a satellite of the Earth; the thing that impressed Galileo was seeing a *system* of satellites. What Galileo saw through his telescope were just the largest and most obvious members of a much more extensive system that now we know to consist of at least 63 satellites, most tiny bodies still not named. The largest satellites are listed in Table 16.1. Those discovered prior to the 1970s were found from Earth-based observations and those discovered in the 1970s from spacecraft images.

The Galilean satellites are all different and all interesting. In 1979, when the spacecraft Voyager I was approaching the innermost Galilean satellite Io, an article by S.J. Peale, P. Cassen and R.T. Reynolds appeared in the journal *Science*, predicting that Io would show active volcanism.[a] The general view was that a body like Io, with mass only 20 per cent greater than that of the Moon, should, like the Moon, have long ago become inactive. To the surprise of almost everyone, this prediction of volcanism turned out to be true. Figure 16.1 shows a volcanic plume from the volcano *Pele* seen by Voyager I. Subsequently many other volcanoes were found on Io.

---

[a] Peale S.J., Cassen P. and Reynolds R.T. (1979), *Science*, **203**, 892.

Table 16.1   The largest satellites of Jupiter.

Name	Mean distance (10^3 kilometres)	Radius (kilometres)	Mass (kilograms)	Year of discovery
Metis	128	20		1979
Adrastea	129	10		1979
Amalthea	181	98		1892
Thebe	222	50		1979
Io	422	1,815	8.94×10^{22}	1610
Europa	671	1,569	4.80×10^{22}	1610
Ganymede	1,070	2,631	1.48×10^{23}	1610
Callisto	1,883	2,400	1.08×10^{23}	1610
Leda	11,094	8		1974
Himalia	11,480	93		1904
Lysithia	11,720	18		1938
Elars	11,737	38		1905
Ananke	21,200	15		1951
Carme	22,600	20		1938
Pasiphae	23,500	25		1908
Sinope	23,700	18		1914

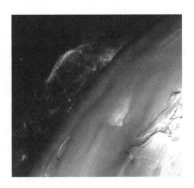

Figure 16.1   The volcanic plume from Pele (NASA).

The basis of the prediction was the relationship of Io's orbit to that of Europa, the adjacent Galilean satellite. Both these satellites are in closely circular orbits with their orbital periods precisely in the ratio 1:2, so that Europa makes one complete orbit of Jupiter while Io makes two orbits. Thus the planets are closest together always at the same points of their orbits, at which points they receive the maximum gravitational nudges from each other. This build up of nudges, always in the same place, makes the orbits slightly non-circular and this is the critical factor. It is at this point we must describe in greater detail the tidal effects that were previously mentioned in relationship to the formation of Saturn's rings.

In Figure 16.2(a), A is an extended body in the presence of a massive body, B, represented as a point mass because we are only interested in its gravitational influence. Since gravitational force depends on distance, the gravitational force per unit mass on material at N, the near-point of A to B, is greater than that at C, the centre of A that, in its turn is greater than that at F, the far-point of A from B. It is a characteristic of orbital motion that there is an inwardly-directed force, in this case the force of gravity acting at C, that prevents the body from flying off in a straight line in whatever direction it happens

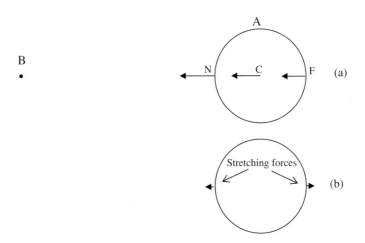

Figure 16.2   (a) Gravitational forces towards B. (b) Stretching forces.

to be moving at the time. From Figure 16.2(a) it will be seen that *relative to* C the force per unit mass at N is towards B while the relative force per unit mass at F is away from B. These differential forces, shown in Figure 16.2(b), stretch the body, which elongates along the line of centres of the two bodies. This stretching force depends on how close body A is to body B so that if the orbit is elliptical then the stretching force varies round the orbit. This alternating greater and lesser stretching injects heat energy into the body A, which we can now identify as Io. This is akin to something that can be observed if a piece of metal is repeatedly bent backwards and forwards. The bending alternately stretches and compresses different parts of the metal, which produces heat by a process known as *hysteresis*. Because Jupiter is so massive, even a slight variation in Io's orbit gives sufficient variation of stretching to provide the energy that drives its volcanoes.

There is another important tidal effect that influences the relationship of a satellite to its parent planet. From Earth, we can only see one face of the Moon because the spin period of the Moon round its axis exactly equals the period of its orbit around the Earth. This is not an accidental relationship but one that comes about because of tidal interactions between the Earth and the Moon. These tidal interactions exist between all planets and their satellites and it is a common characteristic of satellites that they present one face to their parent planets at all times.

Since Io is at a very low temperature, the material that comes from Pele, and Io's other volcanoes, is gaseous sulphur and sulphur compounds. The surface is covered with white and yellow deposits of sulphur dioxide and sulphur (Figure 16.3); the appearance of Io has often been likened to that of a pizza!

Europa, the second of the Galilean satellites going outwards, has two-thirds the mass of the Moon and is covered with a cracked icy surface (Figure 16.4). There are only three craters visible on its surface, suggesting that the ice cannot always have been as solid as it appears to be now. At the temperature of the Galilean satellites, just over 100 K, water ice has the consistency of solid rock rather than being the frangible material that we put into our drinks.

Figure 16.3   The Galilean satellite Io (NASA).

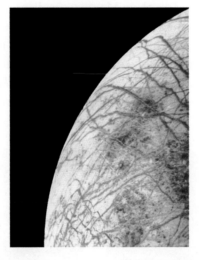

Figure 16.4   The cracked surface of Europa.

The Io–Europa interaction also gives Europa a slightly non-circular orbit although, since Europa is further from Jupiter, the tidal heating within it is much less than for Io. Nevertheless it may be sufficient to melt water below the icy surface so that Europa could have a sub-surface sea of liquid water. Biologists have speculated that some form of life could exist in such a sea and one of the great challenges of some future space mission will be to investigate that possibility.

The next Galilean satellite, Ganymede, is the most massive and largest satellite in the Solar System — its mass is just over twice that of the Moon and its diameter is 10% greater than that of Mercury. However, its density suggests that its bulk structure is about 50% ice and 50% silicate, with perhaps a small iron core. Some parts of Ganymede's surface are heavily cratered and it looks as though all the damage features throughout its history have been preserved in these regions (Figure 16.5(a)). By contrast, other areas show little cratering and are covered by complicated systems of intersecting ridges (Figure 16.5(b)). These ridged areas appear to have undergone considerable surface movements but what caused these features is unknown.

The final Galilean satellite, Callisto, has 1.4 times the mass of the Moon and is slightly less than Mercury in diameter. Its density suggests that, like Ganymede, its composition is an equal mixture of ice and silicates. Callisto's heavily cratered surface must be very old. There are several multi-ringed features on Callisto similar to the Caloris basin on Mercury. The largest of these, *Valhalla* (Figure 16.6), was formed by the collision of a very massive projectile. This must

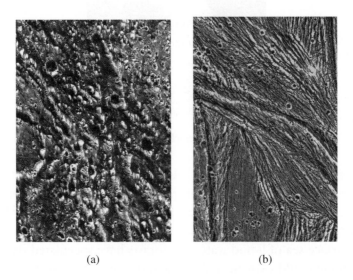

(a)   (b)

Figure 16.5 Ganymede: (a) heavily cratered surface and (b) striated surface.

Figure 16.6   The surface of Callisto showing many craters and the ring system Valhalla.

have happened a long time after Callisto formed because there are few craters in the central basin of Valhalla.

The three innermost Galilean satellites have a remarkable relationship between their orbital periods, which are almost in the ratio 1:2:4. While these ratios are not precise, there is an *exact* relationship linking their rotational speeds (RS) around the planet. This relationship is:

RS of Io + 2 × RS of Ganymede = 3 × times RS of Europa.

This relationship has been explained by an interplay of gravitational and tidal forces involving the three satellites and Jupiter.

The Galilean satellites are *regular satellites* — in direct, close-to-circular orbits in the equatorial plane of the planet. A satellite may have this relationship with its parent planet because of the process by which the satellite was formed. Another reason, especially for a close satellite, is that it is due to gravitational, including tidal, effects. Because of planetary spin, planets tend to bulge out at the equator and the gravitational effect of this bulge on a close satellite is to pull it down towards the planet's equatorial plane. The four satellites closer in than the Galileans in Table 16.1 are all regular according to the definition that has been given, but may be so because of gravitational effects rather than because of their mode of formation.

The other satellites in Table 16.1 fall into two groups of four, at about 11 million kilometres and 22 million kilometres from the planet, respectively. Each group contains many more members; what are listed are just the largest members. These satellites are highly irregular. The orbits of the inner group are inclined between 26° and 29° to the planetary equator and have eccentricities between 0.112 and 0.248. The outer group orbits have eccentricities between 0.114 and 0.244 and inclinations between 145° and 151°, i.e. the orbits are retrograde. Later, in Section 21.2, an idea will be advanced for the origin of these two groups of satellites.

16.2 The Satellites of Saturn

The description of Jupiter's satellites has given a platform for briefer descriptions of other satellites. Table 16.2 gives the largest satellites of Saturn.

The outstanding Saturnian satellite, Titan, is larger than Mercury and second only to Ganymede in mass and size. Its density, 1,880 kilograms per cubic metre, suggests that it consists of a roughly equal mixture of ice and silicates. Its orbital characteristics are those of a regular satellite. Titan is an exceptional satellite in having a thick atmosphere, mainly of nitrogen but with traces of methane and other hydrocarbons. This atmosphere, greater in mass than that of the Earth, is opaque to visible light but radar and infrared imagery shows features interpreted as lakes, seas and craters (Figure 16.7). Because of the prevailing low temperature it is likely that seas and lakes would consist of liquid nitrogen or methane.

Hyperion, an irregularly-shaped satellite with an orbital eccentricity of 0.104, is dynamically linked to Titan with the orbital periods of the two bodies almost precisely in the ratio 4:3.

All the smaller Saturnian satellites known before the space age, recognised in the table by their dates of discovery, have low density and must be predominantly ice. However, since at very low temperatures ice is a rock-like substance, it preserves evidence of damage over the lifetime of the Solar System. For example, Mimas (Figure 16.8)

Table 16.2   The largest satellites of Saturn.

Name	Mean distance (10^3 kilometres)	Radius (kilometres)	Mass (kilograms)	Year of discovery
Pan	134	10		1990
Atlas	138	14		1980
Prometheus	139	46		1980
Pandora	142	46		1980
Epimetheus	151	57		1980
Janus	151	89		1966
Mimas	186	196	3.80×10^{19}	1789
Enceladus	238	260	8.40×10^{19}	1789
Tethys	295	530	7.55×10^{20}	1684
Telesto	295	15		1980
Calypso	295	13		1980
Dione	377	560	1.05×10^{21}	1684
Helene	377	16		1980
Rhea	527	765	2.49×10^{21}	1672
Titan	1,222	2,575	1.35×10^{23}	1655
Hyperion	1,481	143	1.77×10^{19}	1848
Iapetus	3,561	730	1.88×10^{21}	1671
Phoebe	12,952	110	4.00×10^{18}	1898

Figure 16.7   The surface of Titan from combined infrared and radar images taken by the Cassini spacecraft (NASA).

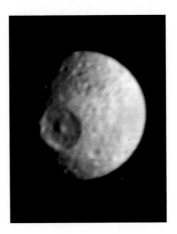

Figure 16.8   Mimas, showing a large impact feature (NASA).

has not only a very cratered surface but also a huge impact feature that must have nearly completely disrupted the satellite.

Enceladus has smooth regions suggesting activity that eradicated surface features after initial heavy bombardment. The pairs of non-adjacent satellites, Mimas-Tethys and Enceladus-Dione, have a precise 1:2 ratio in their orbital periods, similar to that for the Galileans, Io and Europa. These Saturnian satellites may have been affected by tidal heating at some stage in their lifetimes. Their eccentricities are small enough for them to be considered as regular satellites.

There are some satellites in Table 16.2 with identical orbital radii. These are illustrative of a very interesting dynamical situation. In Figure 16.9 we show the relationship of the satellites Tethys, Telesto and Calypso, in the same orbit around Saturn. Tethys is much more massive than the other two, of which Telesto orbits 60° ahead of Tethys and Calypso 60° behind. This arrangement is stable. If for any reason Telesto or Calypso should move away from these positions then they would experience forces pushing them back again. In the language of mechanics, these smaller satellites are said to be at *stable Lagrangian points*. In a similar way Helene is at the leading Lagrangian point of Dione and a tiny satellite, Polydeuces, not in the table, is at the trailing point.

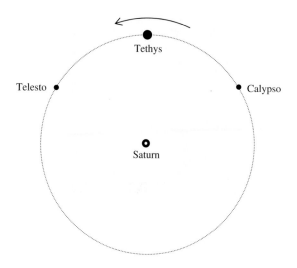

Figure 16.9 The positions of Tethys, Telesto and Calypso in orbit around Saturn.

The satellites Janus and Epimetheus have a different kind of relationship. Although the orbital radii are identical to the accuracy of Table 16.2, they are slightly different. The inner satellite travels faster and slowly approaches the outer one. As they approach, the gravitational forces between them pull backward on the outer one and forwards on the inner one. This pulls the outer one into a smaller orbit and the inner one into a larger orbit so they exchange places. This gavotte takes place every four years and is stable over long periods of time.

Rhea and Iapetus are relatively big satellites, both showing well cratered surfaces. Rhea is regular in its orbital characteristics but Iapetus, with orbital eccentricity 0.028 and inclination to the planet equator 15°, is certainly not regular. Iapetus (Figure 16.10) shows a unique feature; one hemisphere, the one leading in its progress around Saturn, is much darker than the other.

Phoebe is the outermost satellite that, with orbital eccentricity 0.163 and the inclination to the planetary equator 150°, is in a retrograde orbit. It can only sensibly be interpreted as a captured body.

Figure 16.10   Part of the darker hemisphere of Iapetus is seen in the lower left of the image (NASA).

Table 16.3   The largest satellites of Uranus.

Name	Mean distance (10^3 kilometres)	Radius (kilometres)	Mass (kilograms)
Puck	86	85	
Miranda	130	242	7.5×10^{19}
Ariel	191	580	1.4×10^{21}
Umbriel	266	595	1.3×10^{21}
Titania	436	805	3.5×10^{21}
Oberon	583	775	2.9×10^{21}

16.3 The Satellites of Uranus

Uranus has 27 satellites, most of them very small and discovered in the space age. Table 16.3 gives the six largest of these, five known from Earth-based observations and the sixth, Puck, from spacecraft images. They are all regular, orbiting in the equatorial plane of Uranus in a direct sense relative to the curious orientation of the planet's spin axis. They are all of low density and must predominantly consist of icy materials. Figure 16.11 shows the appearance of the smallest satellite, Miranda, the closest to Uranus, which

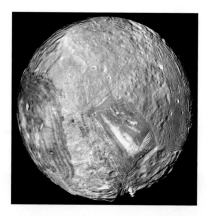

Figure 16.11 Miranda.

shows extensive disturbance of the surface with parallel ridges and troughs.

16.4 The Satellites of Neptune

Before the space age there were two known Neptunian satellites, Triton and Nereid, both with unusual characteristics. Now 13 satellites are known, mostly quite small but one larger than Nereid. The three largest satellites are listed in Table 16.4.

The large satellite, Triton, the seventh most massive in the Solar System, is in retrograde orbit around Neptune. Tidal effects on the retrograde orbit cause Triton to slowly approach Neptune and, at some future time, it will get so close that it will be disrupted by tidal forces. Neptune may then acquire a substantial ring system, rivalling that of Saturn. Triton has a tenuous atmosphere that allows visual observation of its surface. A spectacular view taken by the Voyager II spacecraft in 1989 shows a southern icecap, probably of solid nitrogen (Figure 16.12). Other observations suggest that Triton has volcanoes that emit nitrogen frost containing organic compounds.

The other previously-known satellite, Nereid, is distinguished by its extreme orbit with eccentricity 0.75. Because of its large average distance from Neptune, it can be seen by Earth-based telescopes. The slightly larger Proteus was only seen once spacecraft visited the planet.

Table 16.4   The largest satellites of Neptune.

Name	Mean distance (10^3 kilometres)	Radius (kilometres)	Mass (kilograms)
Proteus	118	209	
Triton	355	1,350	2.14×10^{22}
Nereid	5,509	170	

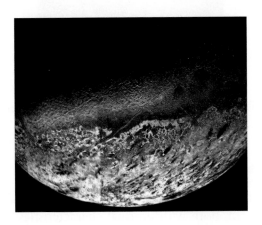

Figure 16.12   A view of Triton showing the southern icecap (NASA).

16.5 The Satellites of Mars

While all the major planets have many satellites and even ring systems, which are almost certainly fragmented satellites, of the terrestrial planets only Mars and Earth have satellite companions. Mars has two satellites, Phobos and Deimos, both small and orbiting closely. Figure 16.13 shows images of these satellites.

Phobos is non-spherical with minimum and maximum diameters of 20 and 28 kilometres. It is covered with craters, the largest of which, Stickney, has a diameter of 10 kilometres. Phobos orbits Mars in a direct sense but its orbital period, 7 hours 40 minutes, is less than the Martian spin period. As seen from the Martian surface Phobos rises in the west and sets in the east. Deimos, also non-spherical, has minimum and maximum diameters of 10 and 16 kilometres. Both

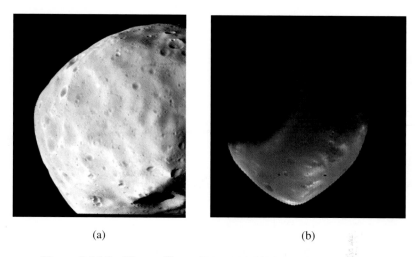

(a)   (b)

Figure 16.13   The satellites of Mars: (a) Phobos and (b) Deimos.

satellites are covered with dust, probably debris from collisions, and the smoother appearance of Deimos suggests that its dust layer is thicker than that on Phobos. These satellites are both irregular and are certainly captured bodies.

16.6 The Moon

The first and obvious feature of the Moon is how large and massive it is compared to its parent planet. Table 16.5 gives the characteristics of the largest seven satellites of the Solar System. The Moon is the fifth largest but is clearly anomalous in the ratio of the planet to satellite mass. The face of the Moon turned towards the Earth is shown in Figure 16.14. The two major types of terrain are heavily cratered highland regions and mare basins, which are lava-plains with generally circular boundaries. The mare basins are caused by large impacts on the lunar surface that subsequently filled with molten basaltic material from below the surface. They are lightly cratered by comparison with the highlands; from the density of cratering, the time of the mare lava flows, i.e. when the volcanism occurred, has been estimated to be between 3.16 and 3.96 billion years ago. Of course, volcanism would have occurred before then, from when the Moon

Table 16.5   Characteristics of the seven largest satellites of the Solar System.

Name	Mass (Moon units)	Radius (kilometres)	Density (kilograms per cubic metre)	Mass ratio (planet:satellite)
Ganymede	2.01	2,631	1.93	12,685
Titan	1.84	2,575	1.88	4,176
Callisto	1.47	2,400	1.47	17,781
Io	1.22	1,815	3.55	21,289
Moon	1.00	1,738	3.34	81
Europa	0.65	1,569	3.04	38,914
Triton	0.29	1,350	2.07	4,785

Figure 16.14   The near side of the Moon.

formed some 4.6 billion years ago, but older material is covered by later eruptions.

It was always assumed that the surface of the Moon as seen from Earth was a fair sample of the whole Moon's surface. In 1959 a Lunik spacecraft launched by the Soviet Union photographed the Moon's far side. The images were of poor quality, but good enough to show, to everyone's surprise, that the rear of the Moon was completely different from the near side. Figure 16.15 is a better quality image of the Moon's rear surface. Most of the surface is of the heavily cratered

Figure 16.15 The rear side of the Moon.

highland type with only one obvious mare feature, *Mare Moscoviense,* seen in the top-left part of the picture. This hemispherical asymmetry of the lunar surface, with one half highland and the other dominantly volcanic, is a feature previously noted in Mars. The first argument that was raised to explain the difference of the two sides of the Moon was that they had different collision histories. It was argued that the gravitational effect of the Earth would focus potential projectiles onto the near face. Theory did not support this argument and it was also shown to be wrong by observations. Lunar orbiters carrying radar equipment have enabled the profile of the lunar surface to be mapped. The far side of the Moon *does* have large basins, as big as those on the near side, but for some reason they were not filled with basalt coming from the interior.

The Apollo series of spacecraft launched by NASA, including missions in which astronauts carried out experiments on, and collected specimens from, the surface, has given a good understanding of the Moon's composition and structure. The highlands consist of igneous rocks and maria material is basalt similar to that coming from terrestrial volcanoes. We also know about the interior of the Moon from seismometers that were left on the surface. The Moon is a quiet body, although there are occasional moonquakes that are extremely weak compared with earthquakes. The total energy of moonquakes is equivalent to the explosive power of 50 grams of TNT per year. The seismometers left on the Moon are very sensitive and can record the tiniest of surface movements including those caused by meteorite falls. In one experiment a third-stage Saturn booster rocket crashed onto the Moon; the energy in this event was equivalent to exploding 1 tonne of TNT. The reverberations went on for more than three hours — compared with a few minutes for an earthquake. The bulk of the Moon behaved like a near-perfectly elastic solid — a comment at the time was that 'it rang like a bell'.[b]

The deduced interior structure shows that the Moon has a small iron core, a mantle of heavier silicates and a crust of lighter rocks. The crust is found to be 48 kilometres thick on the near side of the Moon but 74 kilometres thick on the far side (Figure 16.16). This explains the hemispherical asymmetry of the Moon; projectiles fell uniformly all over the surface but only on the near side were they able to penetrate through to molten material that would fill the resultant basin.

This explanation of hemispherical asymmetry raises an important question — 'why is the crust thinner on the near side?' Early planets

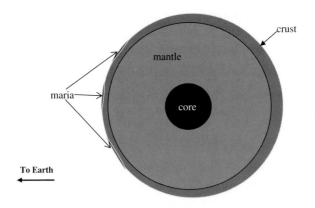

Figure 16.16   A schematic cross section of the Moon (not to scale).

---

[b] NASA, http://science.nasa.gov/science-news/science-at-nasa/2006/15mar_moon quakes (Accessed 10/05/12).

Figure 16.17   A layered fluid satellite in the presence of a tidal field (tidal effects exaggerated for clarity).

and satellites were largely fluid and certainly much less rigid than now. If the Moon formed remote from another body then the crust would have had the same thickness everywhere. However, if it formed close to Earth, or became associated with Earth while still in a pliable state, then the crust should have been *thicker*, not thinner, on the side facing the Earth. Figure 16.17 shows the way that a fluid low-density crust would form under the influence of the tidal effect of a planet. We shall return to this puzzle later to consider a solution.

Chapter 17

'Vermin of the Sky' and Other Small Bodies

17.1 Bode's Law

In 1766, when only planets out to Saturn were known, a German astronomer, Johann Daniel Titius (1729–1796) noticed that the orbital radii of the planets had close to a systematic spacing. This idea was picked up by another German astronomer, Johann Elert Bode (1747–1826), and announced in 1772 without attribution to Titius. This spacing pattern became known as Bode's law and is shown in Table 17.1.

The sequence is based on first adding 0.3 to Mercury's orbital radius to get that of Venus and then repeatedly doubling what is added to Mercury's orbital radius to get the other Bode-law values. The only discrepancy was a gap between Mars and Jupiter, at 2.8 astronomical units, but belief in the validity of the law was strengthened when William Herschel (1738–1822) discovered Uranus in 1781. Astronomers began searching for a planet that they believed had to exist between Mars and Jupiter. On 1st January 1801, the first day of the new century, Giuseppe Piazzi (1746–1826), the Director of the Palermo Observatory in Sicily, discovered a body in the predicted region, which he named Ceres after the Roman patron god of Sicily. Ceres is small, 950 kilometres in diameter (Figure 17.1), but close to the position predicted by Bode's law. When Neptune was

Table 17.1   The orbital radii of the planets and Bode's law.

Planet	Orbital radius (astronomical units)	Bode's law
Mercury	0.387	0.4
Venus	0.723	0.7 = 0.4 + 0.3
Earth	1.000	1.0 = 0.4 + 0.6
Mars	1.524	1.6 = 0.4 + 1.2
(Ceres)	(2.75)	2.8 = 0.4 + 2.4
Jupiter	5.203	5.2 = 0.4 + 4.8
Saturn	9.539	10.0 = 0.4 + 9.6
Uranus	19.19	19.6 = 0.4 + 19.2
Neptune	30.07	38.8 = 0.4 + 38.4

Figure 17.1   Ceres, the first asteroid to be discovered (Hubble).

discovered in 1846, it did not fit Bode's law (see Table 17.1) so, sub-sequently, the law fell somewhat out of favour.

17.2 Asteroids Galore

The astronomical community was content with the discovery of Ceres — it was small but, at the time, it gave a satisfactory complete-ness to the pattern of Bode's law. However, within the next few years three other bodies, Pallas, Juno and Vesta, all smaller than Ceres but still quite substantial, were discovered at similar distances from the Sun. This opened the floodgates to the discovery of large numbers of small bodies, mostly, but not always, in the region between Mars and Jupiter. Tens of thousands of these bodies, called *asteroids,* are now known, with diameters between that of Ceres, the largest asteroid, down to a few kilometres. The numbers of known asteroids became so great that Edmund Weiss (1837–1913), the Director of the Vienna Observatory, called them 'the vermin of the sky', a term subsequently used by other astronomers. They are all in direct orbits but have a wide range of eccentricities and can have orbits inclined to the ecliptic by up to 64°. Table 17.2 lists some of the more interesting asteroids.

The table shows the three large asteroids discovered shortly after the discovery of Ceres. The next two, Hygeia and Undina, are also large and are a little further out, but still completely in the region between Mars and Jupiter. The bigger asteroids tend to be spherical, or nearly so, because the large force of gravity acting on more massive

Table 17.2   A selection of asteroids.

Name	Year of discovery	Semi-major axis (astronomical units)	Eccentricity	Inclination (°)	Diameter (kilometres)
Ceres	1801	2.75	0.079	10.6	950
Pallas	1802	2.77	0.237	34.9	608
Juno	1804	2.67	0.257	13.0	250
Vesta	1807	2.58	0.089	7.1	538
Hygeia	1849	3.15	0.100	3.8	450
Undina	1867	3.20	0.072	9.9	250
Eros	1898	1.46	0.223	10.8	20
Hidalgo	1920	5.81	0.657	42.5	15
Apollo	1932	1.47	0.566	6.4	
Icarus	1949	1.08	0.827	22.9	2
Chiron	1977	13.50	0.378	6.9	

Figure 17.2   The asteroid Eros (NASA).

objects pulls material inwards as far as possible; the shape achieving this goal is a sphere. For smaller bodies the strength of the material of the asteroid can maintain a non-spherical form against gravitational forces. The diameters shown in the table are averages; for example, Eros is an elongated body, 33 kilometres long and 13 kilometres in average width (Figure 17.2). This asteroid has been imaged by a spacecraft from a distance of a few tens of kilometres, showing the fine detail of its surface.

Some asteroids are described as Earth-crossing — at some positions in their orbits they are at one astronomical unit from the Sun. There are about 2,000 such bodies with diameter greater than 1 kilometre; there is a theoretical possibility that at some time one of them could collide with the Earth. That possibility should not cause us to lose sleep — not because such collisions are unknown but rather because they occur at very long time intervals. It has been suggested that the demise of dinosaurs, about 65 million years ago, was the result of an asteroid striking the Earth. An asteroid of diameter 10 kilometres falling on Earth would release energy equivalent to the explosion of 100,000 hydrogen bombs! Apollo, the first Earth-crossing asteroid to be detected, in the year of its discovery, 1932, came within 3 million kilometres of the Earth, just seven to eight times the distance of the Moon. An even closer asteroid passage was on 8th November 2011 when the asteroid 2005 YU55, of diameter 400 metres, came within 325,000 kilometres of the Earth, even closer than the Moon. Another Earth-crossing asteroid in the table, Icarus, has the largest eccentricity of any known asteroid and at its closest is just 0.19 astronomical units from the Sun. Earth-crossing asteroids are known collectively as the *Apollo asteroids.*

There are other asteroids that fall outside the general rule of existing completely between the orbits of Mars and Jupiter. The *Aten asteroids*, with semi-major axes less than one astronomical unit, are mostly small and spend most of their time within the Earth's orbit. Other asteroids, like Eros, stay outside the Earth's orbit but have orbits crossing that of Mars. Chiron has an orbit that is mostly between Saturn and Uranus; there may also be other types of asteroid, too small to be observed, that inhabit other regions of the Solar System. The diameter of Chiron must be at least 100 kilometres for it to be observed at all.

Another interesting group of asteroids, not in Table 17.2, are the *Trojans*, in similar orbits to that of Jupiter but clustered 60° ahead and 60° behind Jupiter in its orbit. This is a very stable arrangement, previously mentioned with respect to some of the satellites of Saturn (Figure 16.9).

17.3 Meteorites

Meteorites are objects, usually small, that fall to Earth and form a valuable sample of extra-terrestrial material. With few exceptions, they are fragments coming from occasional asteroid collisions and, as such, they enable us to determine asteroid compositions. This relationship between asteroids and meteorites is indicated by the way that asteroids reflect light of various wavelengths (colours), which can be matched to the reflectivity of different kinds of meteorite. Every day, something between 100 and 1,000 tonnes of meteoritic material falls onto Earth. This seems an alarming figure but, to put it in perspective, that rate of fall over the whole lifetime of the Solar System would have added a mass about one-tenth of that of the Earth's atmosphere. Meteorites mostly fall into the sea or in remote uninhabited regions, and even when they fall near habitation they are usually unobserved and become unrecognisable as meteorites since they often resemble the normal stones and rocks found on the Earth's surface. The number of meteorite finds has greatly increased in the last few decades as meteoriticists have begun to search for them in places where they would be conspicuous. Rich finds of meteorites are found in the Antarctic. The land of the Antarctic continent is covered by about three kilometres of ice so, if a stony object is found on or near the surface, then one may be sure that it came from above and not from below. Since 1969 there have been many expeditions to collect meteorites in Antarctica, particularly by Japanese and American scientists (Figure 17.3) and more meteorites now come from that source than from all other sources combined.

Larger meteorites often survive passage through the atmosphere although, because of atmospheric friction, surface material may melt and form a dark crust. Frangible meteorites usually disintegrate on passage through the atmosphere and descend as a shower of small objects. Paradoxically, *very* tiny objects may survive the passage to

Figure 17.3   A successful find in Antarctica (NSF, NASA).

Earth almost intact. Because of their large surface-to-volume ratios they radiate heat efficiently and they gently drift down through the atmosphere.

There is much that meteorites can tell us about the origin and evolution of the Solar System. Apart from their physical appearance, which we shall deal with here, the detailed chemistry of meteorites indicates the conditions under which they formed. Another, and very important, aspect of meteorites is that many of them show isotopic anomalies, which is to say that the proportions of different isotopes for some atomic species are different from those measured for terrestrial material. This will be discussed in greater depth in Chapter 22.

There are three main categories of meteorite — stones, consisting mainly of silicates, irons, which are mostly iron with a few per cent nickel, and stony-irons, which are mixtures of the two basic types of material.

17.3.1 *Stony Meteorites*

Stony meteorites occur in three main types, which are:

Chondrites, which contain small glassy spheres called chondrules,
Achondrites, which do not contain chondrules,
Carbonaceous chondrites, which contain volatile materials.

The small glassy spheres in chondritic meteorites were produced from molten rock in the form of a fine spray. Very tiny quantities of any liquid, including water, tend to form spherical droplets because of a physical property called *surface tension*. This applies forces on a drop of liquid to make its surface area as small as possible; the shape having the smallest surface area for a given volume is a sphere. Spherical drops do not form for large volumes of liquid because then the force of gravity overwhelms that due to surface tension. The molten droplets cooled quickly to give the chondrules that then became incorporated in masses of ordinary rocky fragments (Figure 17.4). These subsequently became compacted to form the meteorite material; the largest chondrules in the figure are about a millimetre across.

Figure 17.4   A section through a chondritic meteorite showing many chondrules.

Achondrites contain no chondrules and are similar to some terrestrial and lunar rocks. Rocks can be dated by methods based on the radioactivity of some elements they contain, and most meteorites are found to be about 4.5 billion years old — close to the age of the Solar System. However, a few achondrites are much younger, with ages around one billion years. They contain gas trapped in cavities, and when this gas is analysed it is found to be similar to the composition of the atmosphere of Mars. These meteorites are of three types, *shergottites, nakhlites* and *chassignites,* named after the places where the prototypes were found, and together are referred to as *SNC meteorites.* It is believed that these meteorites originate from the planet Mars, presumably fragments ejected from the surface following an asteroid collision.

Despite their name, carbonaceous chondrites do not always contain chondrules and are mainly distinguished by their chemical compositions. They contain carbon compounds, such as benzene and some amino acids that are the components of proteins (Section 32.2). They also contain water, not in liquid form but as part of *hydrated minerals,* i.e. minerals, such as *serpentine,* that contain water in their structures. Carbonaceous chondrites are usually very dark in colour, sometimes almost black. Unexpectedly, for bodies containing so many readily volatile materials, some of them contain white inclusions of material, known as CAI (Calcium–Aluminium-rich Inclusions), that melt at very high temperatures.

17.3.2 *Iron Meteorites*

Iron meteorites are actually iron-nickel mixtures, with 5–20% of nickel, most of which have cooled down from the liquid state. When the material solidified, it formed two iron-nickel alloys — *taenite* and *kamacite* — the former containing a greater proportion of nickel than the latter. While the material is solid, but still very hot, then the individual atoms have enough energy of motion to migrate around in the solid. This motion of atoms leads to the formation of separate taenite and kamacite regions in the meteorite, giving the characteristic appearance, seen in an etched iron-meteorite cross section in Figure 17.5, known as a Widmanstätten pattern or figure.

Widmanstätten figures give information about the meteorite's thermal history. The longer the meteorite is a hot solid, the larger the needle-shaped platelets of kamacite become. The platelets are small for a quickly cooling iron meteorite and large for one cooling slowly. Once the temperature had fallen to about 600 K then all migration of atoms ceased. Estimates of the cooling rates of iron meteorites are frequently in the range 1–10 K per million years, which indicates the

Figure 17.5   A typical Widmanstätten pattern (NASA).

probable size of the objects from which they came. A football-sized object would cool much faster than that and a large asteroid, the size of Ceres, would cool much more slowly. The estimated size of the source body for most iron meteorites is a few kilometres in diameter, typical of a small asteroid.

17.3.3 *Stony-Iron Meteorites*

Stony-iron meteorites consist of mixtures of iron and stone in approximately equal proportions. There are two main varieties — *pallasites* and *mesosiderites* — their structures indicating different kinds of origin. In *pallasites* (Figure 17.6(a)) silicate crystals are set in a framework of iron. They probably originate from a region of a body where iron and silicate coexisted. *Mesosiderites* (Figure 17.6(b)) contain a chaotic mixture of iron and stone fragments with iron sometimes in globules and sometimes as veins within the stony regions. Some of the stony minerals are of a structure that could not exist at the high pressures in the deep interior of a planet. Mesosiderites can be explained as accumulations of material from violent amalgamations of molten iron-rich and silicate-rich materials.

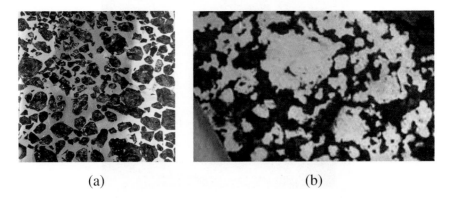

(a)   (b)

Figure 17.6   Cross sections of (a) a pallasite and (b) a mesosiderite (NASA). The iron is lighter coloured.

17.4 Comets

One of the most awesome astronomical sights that can be seen is a near-Earth comet (Figure 17.7). In times past they were regarded as harbingers of great events. In Shakespeare's play *Julius Caesar*, Caesar's wife, Calphurnia, begs her husband not to go to the Senate, referring to recent heavenly portents with the words:

> *'When beggars die there are no comets seen.*
> *The heavens themselves blaze forth the death of princes.'*

The Bayeux tapestry, the record of the invasion of England by William I, shows a comet, Halley's Comet, that appeared in the year 1066. Later, this comet led to an understanding of the place of comets within the Solar System. Edmund Halley (1656–1742), a friend of Isaac Newton, realised that comets were bodies in orbit around the Sun and postulated that the comet seen in 1682 was the one seen in 1607, 1531 and 1456, and he predicted that it would return in 1756. He did not live to see his prediction verified.

From observations of comets it is possible to deduce the characteristics of their orbits. Comets are divided into two categories, *short-period comets* with periods less than 200 years, which spend most of their time within the region occupied by the planets, and *long-period comets*. There are about 100 short-period comets. Those with periods

Figure 17.7   Comet Mrkos, 1957, showing a long plasma tail and a stubby dust tail (Mount Palomar Observatory).

more than 20 years can have any inclination, so their orbits can be either direct or retrograde. Halley's Comet, with period about 76 years, has a retrograde orbit. Comets with periods less than 20 years, called *Jupiter comets*, all have direct orbits, small inclinations of less than 30°, and, for comets, modest eccentricities, usually in the range 0.5–0.7. They are believed to be long-period comets that were influenced by Jupiter, either in several minor interactions or one major one, and were swung into their present orbits. Their aphelia (furthest distances from the Sun) are all about 5 astronomical units, close to Jupiter's orbital radius. A comet loses about one thousandth of its volatile content at each closest approach to the Sun. After about a thousand orbits, taking tens to hundreds of thousands of years, it will have lost all its volatiles and thereafter become an inert dark body. Since short-period comets exist now, although their active lifetimes are so short, there must be some source of new short-period comets to replace those that go through their life cycles.

There is a class of very long-period comets with very extreme orbits that have periods varying from hundreds of thousands to millions of years. Since comets are only seen when they are close to the Sun, these comets must have very large eccentricities, close to unity, the limit for an elliptical orbit. Their aphelia can be at distances of tens of thousands of astronomical units and it has been estimated that there are about 10^{11} (one hundred thousand million) such comets, surrounding the Sun and stretching out half the way to the nearest star. This system of comets is called the *Oort cloud* after Jan Oort (1900–1992), a Dutch astronomer who deduced its existence in 1948. Sometimes a star passing by the Sun will nudge a few Oort cloud comets into orbits taking them close to the Sun and making them visible.

At the heart of a comet is the *nucleus*, a solid object with dimensions of up to a few kilometres. Figure 17.8 shows a spacecraft image of the nucleus of Halley's Comet taken in 1986. The picture shows a peanut-shaped object with bright emanations coming from the left-hand side. The nucleus consists of a silicate framework heavily impregnated with ices of various kinds and these vaporise when the comet approaches the Sun. These vapours create the *coma*, a visible

Figure 17.8   A spacecraft picture of Halley's Comet (ESA/NASA).

sphere round the nucleus of radius between 10^5 and 10^6 kilometres. Outside this, but invisible except to instruments that can detect ultraviolet light, is a vast hydrogen cloud, with ten times the radius of the coma.

A characteristic feature of a comet, seen in Figure 17.7, is its tail or, more precisely, its two tails. When the ices vaporise, they blow off some dusty material and a comet will have two tails, one of gas and one of dust — although they are often so close as to be indistinguishable. It is a popular misconception that the tail points in the opposite direction to that of the motion of the comet, just as the scarf of the driver of an open-top car will stream out in a backwards direction. However, the wind that affects the gas and dust coming from a comet is the *solar wind*, due to charged particles from the Sun moving outwards at several hundred kilometres per second. For this reason the gas tail leaves the comet in an antisolar direction, i.e. it points away from the Sun. The dust is somewhat less affected by the solar wind and so the dust tail is slightly displaced from the gas tail.

17.5 The Kuiper Belt

Our final class of small objects is those forming the *Kuiper Belt*, a swarm of small bodies mostly outside the orbit of Neptune, beyond 30 astronomical units from the Sun and stretching out to an indeterminate distance. It is generally believed that inner Kuiper-Belt objects, perturbed by Neptune, provide the source of new short-period comets. We shall again refer to the Kuiper Belt and what it contains in Chapter 26.

As we have seen, the Solar System is chock-a-block with small objects of different kinds and how they came to be there is a topic to which we shall return.

Chapter 18

Planets Galore

18.1 Detecting Planets Around Other Stars

In 1600 the ill-fated monk Giordano Bruno proposed that there were planets around other stars and that these planets were populated by other races of men. This proposal cost him his life, not so much for making the suggestion but rather for not withdrawing it when put on trial by the Inquisition. Nearly 400 years after his death, his proposal was partly confirmed — there are planets around some other stars — but whether they harbour any form of life is still unknown. Here we shall find out how these planets were detected.

We normally think of planets as orbiting round a stationary Sun, but in reality the Sun is also in motion. Because there are many planets in the Solar System it is difficult to describe the Sun's motion so, to simplify the discussion, we shall consider a star with a single planet. How then can we describe the motion of the two bodies for this simple system?

What really happens is that the star and the planet are *both* in orbit. They both go round a fixed point, the *centre of mass*, which lies on the line connecting them (Figure 18.1). The distance of each from the centre of mass is proportional to the mass of the *other* body; if the star is 100 times as massive as the planet then the planet is 100 times more distant from the centre of mass.

In most cases stars are several hundred times as massive as their planets so, in the figure, the stellar orbit has been exaggerated to make the presentation clearer. If the star were 1,000 times as massive

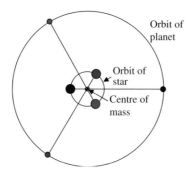

Figure 18.1   The orbits of a star and its planet.

as the planet then the orbit of the star would be 1,000 times smaller than that of the planet and, since the orbits have the same period, the orbital speed of the star would be 1,000 times smaller than that of the planet. Jupiter has three times the combined mass of all the other planets and one thousandth of the mass of the Sun so, ignoring all planets other than Jupiter, this is approximately the situation in the Solar System. The speed of Jupiter in its orbit is 12 kilometres per second, so the speed of the Sun in its orbit is one thousandth of that, i.e. 12 metres per second.

Even with powerful telescopes it is not generally possible to produce an image of a planet since the light from the star would overwhelm the light reflected off the planet. We use the word 'generally' because, since 2008, images of planets *have* been obtained (Section 18.3). However, if the line of sight from the Earth were in the plane of the star's orbit, the star would first retreat from and then approach the Earth in a periodic fashion. If this motion could be measured then, although we could not see the planet directly, we could see a consequence of its presence. A means of detecting and measuring the speed of a light-emitting object along the line of sight is the Doppler Effect. Detecting the speed of a star due to an orbiting planet, with speeds just a few metres per second, presents a considerable challenge to optical measuring techniques. As an example, the speed of the Sun around the centre of mass due to Jupiter's orbit, 12 metres per second, is a fraction 4×10^{-8} of the speed of light. Thus if a spectral line in a stellar spectrum had a wavelength of 5×10^{-7} metres then the shift in that wavelength due to a speed of 12 metres per second is $(4 \times 10^{-8}) \times (5 \times 10^{-7}) = 2 \times 10^{-14}$ metres. This would seem to be an impossible shift to measure but, by the most advanced application of an optical technique known as interferometry, such changes, and even smaller ones, can be measured. There are some complications in these measurements since the Earth itself is moving round the Sun and the centre of mass of the star-planet system may also be moving relative to the Sun but allowance can be made for these factors and the speed of the star in its orbit determined.

Figure 18.2 shows early observations of variable stellar radial velocities taken over a ten-year period for the star 47 Uma. The difference of the maximum and minimum of the fitted curve, which have some errors, gives twice the speed of the star in its orbit. The period of the fluctuation, 1,094 days, gives the period of the stellar, and hence also of the planetary, orbit.

When we observe a system consisting of a planet orbiting a star, the quantities we wish to find are the masses of the star and the planet and the orbital characteristics — its semi-major axis and eccentricity. Knowing just the orbital speed of the star and its orbital period — the information obtained from Figure 18.2 — is insufficient to estimate

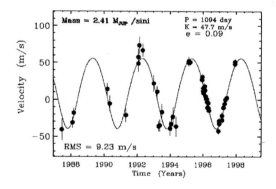

Figure 18.2 Stellar velocity measurements for 47 Uma.[a]

---

[a] Mayor, M. and Queloz, D. (1995), *Nature* **378**, 355.

the mass and orbit of the planet and the mass of the star. Fortunately, we have a way of estimating the mass of the star independently of the presence, or absence, of a planetary companion. In Chapter 10 we showed that for main-sequence stars their temperatures and masses are related. If the temperature of a main-sequence star is known — something that can be found even for very distant stars — then its mass can be found. Most stars that have been found to possess planets are main-sequence stars and so their masses can be estimated.

The theory of planetary motion developed by Isaac Newton connects the mass of the star, and the period and semi-major axis of an orbiting planet; if any two of these three quantities are known then the third can be found. Since observations give the mass of the star and the period of the orbit then the semi-major axis can be found. Assuming a circular orbit for now, the semi-major axis is the radius of the orbit and gives the total distance travelled by the planet in one period and this gives the speed of the planet in its orbit. But the ratio of the speed of the star to the speed of the planet, both now determined, is the ratio of the mass of the planet to the known mass of the star so that the mass of the planet can be found.

The above discussion assumes that the star-planet orbital plane is in the line of sight so that the motion of the star at the edges of the orbit is directly towards or away from the observer. This assumption is generally untrue; Doppler-shift measurements give the speed *along the line of sight*, which will usually be smaller than the speed of the star. The relationship between the speed along the line of sight and the true speed of the star is shown in Figure 18.3.

The calculation shows that the mass of the planet is proportional to the speed of the star. Hence, if the speed of the star is

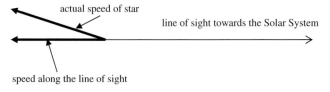

Figure 18.3   The speed along the line of sight is smaller than the speed of the star.

underestimated then so is the mass of the planet. Consequently only a *minimum* planetary mass can be found and the true mass of the planet will be greater by some unknown factor.

There is a rare situation when a true planetary mass can be estimated. This is when the line of sight is so close to the orbital plane of the planet that the planet moves across the disk of the star. Neither the planet nor the disk of the star can be seen; the planet's passage across the disk is indicated by a diminution of the light as the planet moves across. An excellent example of such an observation is shown in Figure 18.4. The proportion of reduction in the brightness of the star gives the ratio of the cross-sectional area of the planet to that of the star. It happens that, just as the masses of main-sequence stars are

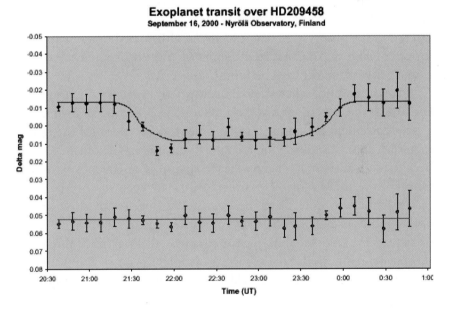

Exoplanet transit over HD209458
September 16, 2000 - Nyrölä Observatory, Finland

Figure 18.4 The light curve from the transit of an exoplanet over the star HD 209458. The lower line is from a check star. This observation was made by amateur astronomers in Finland at the Nyrölä observatory.[b]

[b] URSA, http://www.ursa.fi/yhd/sirius/HD209458/HD209458_eng.html (Accessed 10/05/12).

known from their temperatures, so are their radii and, in this way, it is possible to estimate the radius of the planet as well as its true mass.

In December 2011 NASA announced the discovery of a planet, Kepler 22b, orbiting a star about 600 light years from the Sun.[c] The planet transits the disk of the star, giving an estimated diameter about 2.4 times that of the Earth. From the orbital period, 290 days, and the mass of the star, 0.97 solar mass, the distance of the planet from the star is found to be about 0.85 astronomical units. That distance, taken in conjunction with the luminosity of the star, which is some-what less than that of the Sun, gives a surface temperature of 22° C. The mass of the planet has not been found, but if it gave a density similar to that of the Earth then it could contain oceans and conti-nents and some forms of life.

So far the assumption has been made that the orbits are circular but in general they will be ellipses. For an elliptical orbit the curve fitted to the observed stellar speeds is a distorted version of that shown in Figure 18.2 and the extent of the distortion enables the eccentricity of the orbit to be found.

Planets around other stars are known as *exoplanets.* During the late 1980s there were tentative claims for the detection of exoplanets but measurement techniques at that time were too poor for the claims to be accepted. Since 1995 reliable discoveries of exoplanets have been made and there has been a steady stream of discoveries since that time; by the middle of 2011 over 1,200 had been discovered. The easiest planets to detect are those with large masses in close orbits, both conditions making the speed of the star greater and hence easier to detect by Doppler-shift measurements. A close orbit also gives shorter orbital periods so enabling a complete cycle of the planetary motion to be found more quickly. Improvements in interferometry techniques have lowered the threshold of the lowest planetary mass that can be estimated, so that planets with masses down to about one half of those of Uranus and Neptune can be found if the planets are in favourable orbits. Apart from the problem of giving low stellar

──────────

[c]Borucki W.J. *et al.* (2011), *Earth and Planetary Astrophysics*, 21.

speeds, large orbits introduce another problem. For a planet with a star and orbit like those of Neptune the period would be more than 160 years, so it would take several tens of years just to detect a change in the speed of the planet along the line of sight. For this reason we cannot estimate with any degree of certainly what proportion of stars has planetary companions. It is probably safe to say that *at least* 7% of stars do so but it is possible that the proportion is much larger.

The characteristics of a number of exoplanets are given in Table 18.1. There are some exoplanets with very close orbits. Mercury's orbit has a semi-major axis of 0.4 astronomical units but the planets in the first four entries in the table are at about one sixth to one tenth of that distance from their stars. We also notice that υ-Andromedae has a *family* of planets, at least the three that have been detected. When there are, say, three planets then the star velocity values have to be fitted to a complicated curve which is the sum of three simple curves with different periods and different amplitudes.[d]

Table 18.1 Characteristics of a sample of exoplanets.

Star	Minimum mass of planet (Jupiter units)	Period (days)	Semi-major axis (astronomical units)	Eccentricity
HD 187123	0.52	3.097	0.042	0.03
τ-Bootis	3.87	3.313	0.0462	0.018
51 Peg	0.47	4.229	0.05	0.0
υ-Andromedae	0.71	4.62	0.059	0.034
	2.11	241.2	0.83	0.18
	4.61	1266	2.50	0.41
HD 168443	5.04	57.9	0.277	0.54
16 CygB	1.5	804	1.70	0.67
47 Uma	2.41	1,094	2.10	0.096
14 Her	3.3	1619	2.5	0.354

[d]The amplitude of the curve is one half of the difference between the maximum and the minimum.

A final point of interest is the high eccentricity of some orbits, much higher than that of Mercury, 0.206, the highest for any planet in the Solar System. The highest eccentricity in the table is 0.67 and there are some exoplanets with even higher values. There are also examples of combinations of very close and very eccentric orbits.

For more than 200 years scientists have been proposing scientifically-based theories for the origin of the Solar System. Logically, that goal should now be changed to that of finding a theory for the formation of planetary systems in general. Then, once a plausible theory for the origin of planets has been found, further consideration can be given to discovering mechanisms that would give the detailed structure of our own planetary system.

18.2 Imaging the Effects of Unseen Exoplanets

Substantial disks of gas and dust are found around young stars, and even the Sun, by no means young, possesses a dusty disk. This can be seen by the sunlight it scatters and appears as a band of light in the sky, called the *zodiacal* light (Figure 18.5), concentrated towards the ecliptic. It is best seen either just after sunset or just before sunrise, when the sky is dark but the Sun is not too far below the horizon. The zodiacal light is seen when looking in the direction of the Sun and is due to dust scattering in a forward direction. Looking away from the Sun, a much fainter band, the *gegenschein* or *counterglow*, can sometimes be seen due to backscattered light from dust outside the Earth's orbit.

Small bodies, such as dust grains, spiral in towards the Sun due to the solar radiation that falls on them; this phenomenon is known as the *Poynting–Robertson effect*. A grain of sand in the Earth's orbit would join the Sun after one million years or so, which tells us something important — that there must be a constant source of dust in the Solar System that could maintain its dust disk over its lifetime. In the Solar System there are two important sources: grains shed by a comet's dust tail and occasional collisions between asteroids that produce fine dust as well as meteorites.

In principle the Earth's dust disk could be detected from a distant star by the infrared radiation it emits, but in practice this radiation

Figure 18.5 The zodiacal light (NASA, courtesy of nasaimages.org).

would be of such low intensity that it would be virtually undetectable. However, the disks around young stars are more than one thousand times denser that that of the Sun and hence emit enough infrared radiation for us to be able to detect them.

There are some stars, not very young but considerably younger than the Sun, which have substantial detectable dusty disks and hence must contain some rich sources of dust. In recent years it has been possible to image these disks with infrared radiation with a resolution sufficient to study their detailed structures. The star Fomalhaut, with 2.1 times the solar mass, is at distance 25 light years. Its dusty disk is clear in the central region, suggesting the presence of planets sweeping up the dust. There is also a bright ring of radius 40 astronomical units, which may be a region rich in comet-like bodies. A star of similar mass and distance, Vega, also has a cleared-out centre and also a bright, concentrated infrared source at 80 astronomical units from the star. The latter feature is thought to be a dusty cloud surrounding a major planet. A planet at that distance from its star would not be detectable by Doppler-shift observations.

The star β-Pictoris, of mass 1.75 $M_\odot$[c] and distance 63 light years was the first star to have its dust disk imaged. The disk has bright rings at 14, 28, 52 and 82 astronomical units from the star. It has been suggested that the spaces between the rings have been swept clean either by the direct passage of a planet or by resonance effects whereby regions corresponding to an orbital period with a simple relationship to that of a planet (e.g. 1/2, 2/3, 3/8) are cleared of debris.

A final star we mention in this context is ε-Eridani, of mass $M_\odot$ and a distance of 10.5 light years. There have been claims of the detection of a planet around this star by Doppler-shift measurements but the measurements are not good enough to reliably support the claim. There is a belt-like structure in the dust disk of a similar kind to that around β-Pictoris.

These infrared observations indicate the probable presence of planets but give less information about the nature and orbits of those planets than do Doppler-shift observations. What they do indicate, however, is the probable presence of planets at distances from stars considerably larger than that of Neptune from the Sun.

18.3 Images of Exoplanets

In 2008 the American astronomer Paul Kalas (b. 1967) produced the first image of an exoplanet. A sharp inner boundary of the dust disk of Fomalhaut indicated the presence of a planet, and detailed examination of images of the dust disk taken in 2004 and 2006 showed the

---

[c] The symbol for the Sun is $\odot$, so $M_\odot$ is a solar mass.

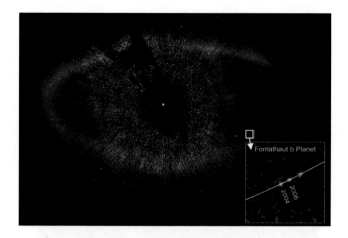

Figure 18.6 The dust disk around Fomalhaut. The enlarged image in the box shows the positions of the planet in 2004 and 2006 (NASA/ESA and P. Kalas, University of California, Berkeley).

planet, which had moved in the interval (Figure 18.6). It is not possible from existing information to estimate the mass or orbit of this planet but it is situated at 115 astronomical units from the star and its orbit is probably not too far from circular.

Also in 2008 an image was released by the Gemini Observatory of three planets around HR 8799, a star of 1.5 solar masses in the constellation Pegasus. These planets, of masses estimated as 10 M_J, 10 M_J and 8 M_J^f, are at distances 25, 40 and 70 astronomical units from the star respectively.

Distortions in the disk of β-Pictoris indicated that there ought to be a major planet orbiting the star at a distance of around 10 astronomical units. At the end of 2008 a team of French astronomers produced a picture that seems to show a planet at a distance of about 8 astronomical units (Figure 18.7). It was not a straightforward image as for Fomalhaut and HR 8799. The outer part of the image shows the reflected light from the dust disk. The inner circular region was imaged in the near infrared, at a wavelength of 3.6 microns. In

---

[f] M_J is the mass of Jupiter.

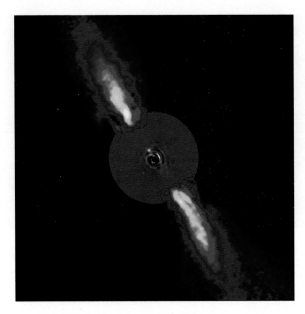

Figure 18.7   A composite image of β-Pictoris. The planet is the bright spot near the centre. (European Southern Observatory.)

the circular region the contribution of the glare of the star was artificially subtracted and what was left was the bright spot indicating the presence of the planet, at 8 astronomical units from the star and with an estimated mass of 8 M_J.

Before 1995 it was strongly suspected that there were planets around other stars but it had not been conclusively demonstrated. Now there is no doubt of their existence — we can sometimes estimate their masses and orbits and in some cases even image them directly.

Forming the Solar System

Chapter 19

Making Planets

19.1 Embedded Clusters, Stars and Protostars

In Chapter 13 some of the consequences of interactions between stars within an embedded cluster were described. We now return to the theme of the development of the Universe; having discussed the formation of stars, next in the hierarchy of producing bodies of ever decreasing size are planets. The previous five chapters have established a factual background for planets in general and the Solar System in particular.

There is no universally-accepted theory for planetary formation. The theory which, by virtue of the number of people working on it and the mass of literature describing it, has the status of being the 'standard theory' describes the formation of a star and planets from the same nebula, a large cloud of dusty gas. However, this Solar Nebula Theory is not in a well-defined state and has many problems, something acknowledged by those who work on it. There is not even agreement about the actual process by which planets are formed — all proposed mechanisms have difficulties. An alternative model, the Capture Theory, described here, has no obvious difficulties and has both observational and theoretical support for its basic assumptions.

In an embedded cluster there are many stars either in or close to the main-sequence stage so that they are quite compact objects. The Sun has a radius of 700,000 kilometres and even a developing star with five or six times that radius could still be regarded as compact. There are also protostars present at every stage of development

215

from being newly-compressed material to just before the radiation-controlled descent to the main sequence. By the time the compressed material has become a recognisable spherical protostar it has a density somewhere in the range 10^{-14} to 10^{-13} kilograms per cubic metre, which corresponds to a free-fall time of between 7,000 and 20,000 years. Because free fall begins so slowly, for most of this time the protostar is an extended low-density object. If the total mass of the protostar were, say, one-half a solar mass with a density 3×10^{-14} kilograms per cubic metre then its radius would be 2×10^{11} kilometres or 1,300 astronomical units. We now consider an object of that size, moving in a not-very-dense embedded cluster in which compact stars are less than 14,000 astronomical units apart. Figure 19.1 gives a general impression in two dimensions of the relative size of the protostar compared to the separation of stars.

The average speed of stars and protostars within an embedded cluster is estimated to be in the range 0.5 to 2 kilometres per second. During the, say, 15,000 years that a protostar is in a very extended state, moving at a speed of 1 kilometre per second it will travel more than 3,000 astronomical units and have a significant probability of interacting with a compact star. The idea that interactions are common between stars and that collisions can occur is the basis of the ideas, discussed in Section 13.2, about how binary star numbers are affected and how massive stars are formed in an embedded-cluster environment.

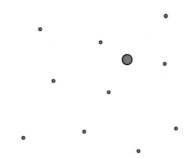

Figure 19.1   An impression of the size of a protostar (larger circle) compared to the separation of stars in an embedded cluster with star number density 100 per cubic light year.

19.2 Interactions between Stars and Compressed Regions

The type of interaction suggested by the above discussion is that between an extended protostar and a compact star. However, before we analyse that possibility in more detail we first consider the process that forms protostars. The initial stage of protostar formation, as described in Section 12.3, is the collision of turbulent elements that produce a compressed region. This will cool before it has greatly re-expanded and, if it then exceeds the Jeans critical mass, it will begin to collapse. The initial density of the compressed region will be less than the 10^{-14} kilograms per cubic metre taken as the minimum initial density of a protostar. However, not all compressed regions will form a protostar, destined to collapse into a star. If the compressed region is less than a Jeans critical mass, or the gas streams collide so violently that the material is either compressed into a pancake form that is not conducive to gravitational collapse or is violently spattered sideways and dispersed, then a protostar cannot form. Even if a compressed region begins to form a protostar, it can be dispersed by further collisions of turbulent gas streams. For these reasons the number of compressed regions being formed should considerably exceed the number of protostars formed and hence interactions between compact stars and compressed regions will be much more common than those involving protostars.

We first describe a simulation of a tidal interaction between a compact star and a compressed region. The simulation uses a computational technique called *smoothed-particle hydrodynamics* (SPH), designed in the late 1970s to model the behaviour of a fluid — in this case a gas — and widely used in astronomical simulations. In a real situation involving gaseous material there are forces due to gravity, pressure and viscosity, the last of which is the resistance a fluid has to flowing (treacle has a high viscosity and water a low viscosity). In SPH the fluid is represented by a set of particles with the properties of mass, velocity and heat content, and forces between the particles simulate the forces that occur in nature.

Figure 19.2 shows a simulation as a series of arrangements of SPH points at various times. The simulations are three-dimensional but the

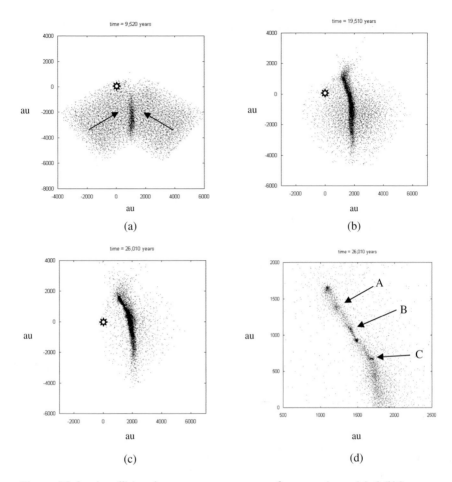

Figure 19.2   A collision between two streams of gas at times (a) 9,520 years, (b) 19,510 years and (c) 26,010 years. Frame (d) shows a higher-resolution view at 26,010 years.

projections shown give a good impression of what is happening. The star is of solar mass. Frame (a) shows two streams of gas, each with half a solar mass and a density 4×10^{-15} kilograms per cubic metre, forming a higher-density region where they collide. The speed of each stream is 1 kilometre per second, and their directions of motion give the compressed region moving around the star in an anticlockwise direction. In frame (b) the compressed region has grown, is being stretched out in the form of a filament and is moving around the star. In frame (c) the filament is beginning to break up into blobs, in the way predicted by James Jeans and illustrated in Figure 8.3. Frame (d) shows a higher-resolution view of the top end of the filament in frame (c), which shows the formation of five distinct condensations. The three condensations marked A, B and C are captured into elliptical orbits around the star. Their masses, in Jupiter mass units (M_J), are: A (1.00), B (1.6) and C (0.75). Their final orbits around the star are all very extended, with semi-major axes of order 1,000 astronomical units (au) and with high eccentricities in the 0.8 to 0.9 range. We shall have more to say about that later.

All five condensations collapse to form planetary-mass objects but two are not captured and escape into the general body of the cluster. In 2000 two British astronomers, P.W. Lucas and P.F. Roche, were searching the Orion Nebula for brown dwarfs, objects with masses intermediate between those of planets and stars — in the range 13–70 M_J — by detecting their infrared radiation.[a] For a mass above 13 M_J the body becomes hot enough in its interior for nuclear reactions involving lithium and deuterium, the isotope of hydrogen, to take place. Above 70 M_J the temperature becomes high enough for reactions involving hydrogen to occur and the body is then defined as a main-sequence star. Lucas and Roche found 13 objects that clearly had masses below 13 M_J and they called them *free-floating planets*. Since then many more similar bodies have been found. Some astronomers who discovered these bodies commented on the fact that these objects provide yet another challenge for the Solar Nebula Theory, since that theory forms stars and planets from a single mass of material and there is no obvious way for the planets to break free. However, free-floating planets occur quite naturally as a consequence of the capture-theory mechanism.

The capture-theory mechanism is extremely robust and planets, or heavier condensations similar to brown dwarfs, form over a wide range of conditions. To illustrate this, another interaction is portrayed in Figure 19.3, where again the mass of the star is a solar mass. The

---

[a] Lucas P. and Roche P. (2000), *Mon. Not. R. Astron. Soc.*, **314**, 858.

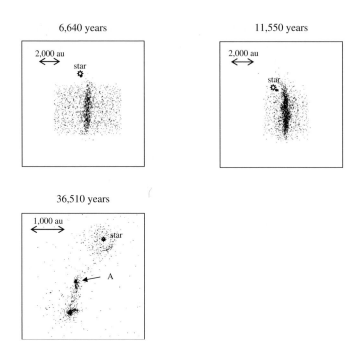

Figure 19.3   A capture-theory interaction leading to a captured brown dwarf, an escaping brown dwarf and an annular ring of captured material.

speed of each gas stream is, as before, 1 kilometre per second but the streams are now of solar mass and they are colliding more head-on than was the case for the Figure 19.2 collision. Two large condensations are formed. The one marked A is captured and has a mass 20 M_J, which puts it into the brown-dwarf category. The other condensation, which is an even more massive brown dwarf, is released into the cluster as a free-floating object. A feature of the Figure 19.2 interaction, which is not very clear in the illustration, is that a considerable quantity of the colliding streams is captured and forms a disk around the star of mass somewhere in the range 50 to 60 M_J. This capture of material is more evident in Figure 19.3, which shows that, in this case, the material forms a doughnut-like structure around the star. This pattern of captured material is not the most common form and we shall later find that it may be influential in producing some of the more unusual orbits found for exoplanets.

19.3 Interactions between Stars and Protostars — How Many Planetary Systems?

Although capture interactions take place more frequently with dense gas regions than with fully-formed protostars, the latter bodies still readily give planet formation. In Figure 19.4 the mass of the protoplanet is 0.35 times that of the Sun and it has a radius of 800 astronomical units. The compact star has the mass of the Sun. The chain of five smaller condensations in frame (d) are all captured and have

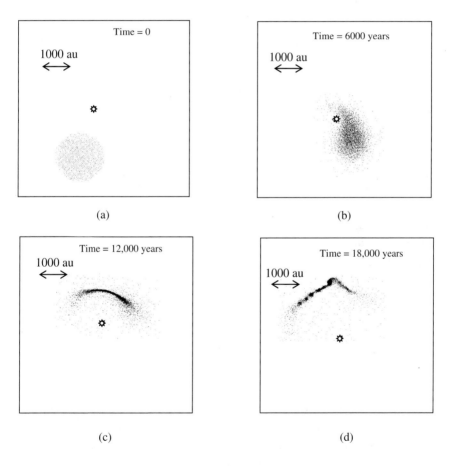

Figure 19.4 The interaction of a protoplanet with a compact star giving captured planets.

masses, in M_J, starting from the left, 4.7, 7.0, 4.8, 6.5 and 20.5, the last being in the brown-dwarf range. It is evident from Table 18.1, which gives *minimum masses*, that planets of more than Jupiter mass are commonly observed. Most, but not all, of the condensation masses from the simulations in Figures 19.2, 19.3 and 19.4 are substantially more than a Jupiter mass but, as we shall find in Chapter 21, not all the mass of a condensation ends up in the final planet.

Large numbers of calculations, with wide variations of the star–protostar interaction conditions, show that the mechanism is extremely robust in producing planets with a large range of masses, some captured by the star and others released as free-floating planets. From this experience a model was set up to give an approximate estimate of the number of stars that might be expected to possess planetary systems. The following were the characteristics of the model:

(i)   A number density of stars in an embedded cluster is randomly chosen in the range 300–3,000 per cubic light year.

(ii)   A spherical region of radius R is found that will accommodate 1,000 stars with the chosen number density.

(iii)   The 1,000 stars are randomly placed within the sphere.

(iv)   Each star is given a random speed in the range 500–2,000 metres per second in a random direction.

(v)   Each star is assigned a random mass in the range 0.8–2.0 solar masses.

(vi)   The protostar is placed at the centre of the sphere with a velocity chosen as in (iv).

(vii)   The mass of the protostar is chosen randomly in the range 0.25–0.75 solar masses.

(viii) For a temperature of 20 K and a hydrogen-helium composition the density and radius of the protostar are found on the basis that it forms a Jeans critical mass.

(ix)   The free-fall time, t_{ff}, is found and the assumption is made that the protostar is capable of producing a filament for a time $t_i = \frac{1}{2} t_{ff}$.

(x)   The motions of the stars and protostar are computed over a time t_i and the closest approach of the protostar to a star is recorded as a fraction of R.

(xi)   Steps (i) to (x) are repeated 1,000 times.

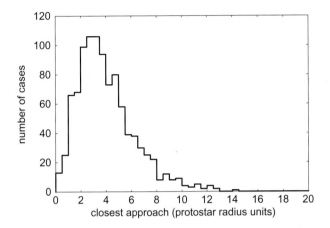

Figure 19.5 The histogram for closest approach distances for 1,000 protostars.

The results of this exercise are illustrated in Figure 19.5, which shows a histogram — a block diagram — giving the number of approaches in intervals of $0.5R$.

Experience with large numbers of capture-theory calculations indicates that a closest approach in the interval 0.5–1.5 R will almost certainly give the protostar stretched into a filament and planet formation. The proportion of cases meeting this criterion in Figure 19.5 is 0.092 — an indication of the proportion of proto-planets that would produce planets round a star. However, that is not precisely what we wish to find, which is the proportion of *stars* with planets. The statistics of numbers of stars with different masses shows that there are five times as many stars with masses 0.25–0.75 solar masses, the range chosen for the protostars, as there are stars of 0.8–2.0 solar masses, the range chosen for the compact stars. Assuming that no star has more than one capture-theory interaction this would give the number of stars with planets as $0.092 \times 5 = 0.46$, nearly one half.

This is not a precise analysis but it does indicate that a high proportion of main-sequence stars, perhaps of the order of 50% of them, should have accompanying planets, especially adding the contribution of interactions with dense regions, which have not been included.

19.4 What Problems Remain?

Embedded clusters with 1,000 stars per cubic light year correspond to a fairly high number density, although higher number densities are found. It appears that interactions of stars just with protostars would comfortably satisfy the 0.07 minimum proportion suggested by observations. However, given that interactions with dense compressed regions also occur, the indication that a high proportion of main-sequence stars, of order 50%, should possess planetary companions would comfortably be met. Nevertheless, before assuming that the theory gives a satisfactory outcome there is another consideration to take into account. A dense embedded environment may be favourable for forming planetary systems but it is also conducive to breaking them up again. This is a matter to which we shall return in the next chapter.

We have established that, in the crowded environment of an embedded cluster, planetary condensations can form by interactions between a compact star and some other body. This other body can be in any form between being a compressed region formed by the collision of turbulent streams of gas to being a well-formed protostar. Planet formation thus appears to be an inevitable outcome of an environment which is crowded with many well-developed stars and within which new stars are forming. It must be stressed that this environment is one that is confirmed by observation, not one that has just been postulated for the *ad hoc* purpose of providing a theory of planetary formation.

Although a promising beginning for a theory of planet formation has been given, there is far to go before demonstrating that there is a final outcome as observed for exoplanets and the Solar System. The next question to address is that of the orbits; all the simulations shown here, and other simulations published in scientific papers, give initial orbits with semi-major axes mostly between 1,000 and 3,000 astronomical units and eccentricities mostly between 0.5 and 0.9. It must be shown that these initial orbits can, in some way, evolve into the kinds of orbits that are actually observed.

Chapter 20

Shrinking Orbits and the Survival of Planetary Systems

Planetary orbits produced by the capture-theory mechanism, with semi-major axes of 1,000 astronomical units or more and high eccentricity, are very unlike those of the Solar System or those deduced for exoplanets. The outermost solar-system planet, Neptune, is 30 astronomical units from the Sun and detected exoplanets are, at most, a hundred or so astronomical units from their parent stars. To understand the extent of the problem of transferring from the orbits found by the Capture Theory to those presently observed, Figure 20.1 shows an orbit with semi-major axis 1,000 astronomical units and eccentricity 0.9 compared to the orbit of Neptune, which is almost perfectly circular. For the capture-theory mechanism to be plausible there must be some process for producing the required decay (reduction of size) and rounding-off, (reduction of eccentricity) of orbits from the initial to the final states.

20.1 Resistance and Decaying Orbits

Planet formation, as shown in Figures 19.2, 19.3 and 19.4, gave rise not only to retained and free-floating planets but also to a substantial amount of captured material forming a disk around the star. This material constitutes a resisting medium within which the newly-formed planet moves. Resistance to motion causes the planet to lose energy, giving an orbit of ever decreasing size. A smaller-scale example

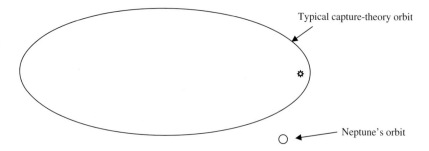

Figure 20.1   A comparison of an initial capture-theory planetary orbit and that of Neptune.

of this is what happens to the orbits of artificial Earth satellites. The atmosphere is very thin at heights of tens of kilometres above the Earth but, nevertheless, it causes the gradual decay of the satellite's orbit so bringing it closer to the Earth. The closer it gets to the Earth the denser the atmosphere is and the faster the orbit decays. The final stage of orbital decay is quite rapid and difficult to predict accurately, so there is always concern about where it will land — the sea or somewhere in an unpopulated land area being desirable.

The resistance of a fluid to the motion of a body within it is a matter of experience — a cyclist feels the resistance of the air on his body when cycling quickly, even on a windless day. Cars are designed to reduce the drag of the air to improve their efficiency in terms of fuel consumption. In particular, liquids exert strong resistance to the motion of bodies through them. A hand quickly moved in water experiences resistance, and the faster the hand moves the greater is the resistance. That is the way that resistance operates. A stationary object within a resisting medium experiences no force. When it moves it will experience a force in a direction that opposes the motion and the greater the speed relative to the medium the greater is the force.

There are several different mechanisms giving resistance forces. The one experienced in everyday life applies to objects of any kind moving in any fluid and depends on the nature of the resisting medium and the size and shape of the moving object. Another mechanism, applicable in astronomical contexts, is where the mass of the moving object, acting gravitationally on the medium, plays a role.

Finally there is a mechanism where both the gravitational action of the object on the medium and that of the medium on the object are important.

20.2 Viscosity

The everyday kind of resistance depends on a property of a fluid called *viscosity*, already mentioned in regard to SPH (see Section 19.2). Viscosity depends on the resistance of a liquid to flowing; water flows easily and has a low viscosity, treacle flows less readily and has a higher viscosity, and pitch, which becomes more fluid on a hot day, flows extremely sluggishly and has a very high viscosity. Normal window glass is a fluid of *extremely* high viscosity, so high that for all practical purposes it behaves like a solid. Nevertheless its fluid nature can be detected; if window glass in an old building is examined it is found to be thicker at the bottom than at the top — it flows downwards at an imperceptible rate.

Viscosity is due to internal friction that inhibits the relative motion of neighbouring layers of fluid. A body moving through a fluid drags the fluid in its immediate neighbourhood with it, while distant fluid is little affected. This causes relative motion of various layers of the fluid and frictional forces occur between the layers. These viscosity forces within the fluid react back onto the object causing the motion, and this reaction constitutes a resisting force on the moving object.

A way of reducing fluid resistance is to streamline the body that moves through it. Car and aircraft manufacturers streamline their products and even the helmets of racing cyclists are designed to give streamlined flow. Figure 20.2 shows air flow around a streamlined object. The flow is smooth and consequently the viscosity force on the object is small. A blunt object, for example a cube, would lead to chaotic motion of the fluid in its vicinity, something called turbulence. With turbulence present, neighbouring regions of fluid move rapidly with respect to each other, which generates big viscosity forces and hence high resistance to the motion of the body.

For bodies of similar shape and density, viscous resistance affects a smaller body more than a larger body. The viscous force depends on

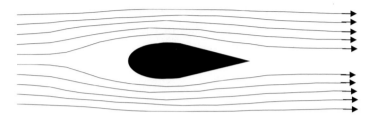

Figure 20.2   The motion of a fluid around a streamlined body.

the area impacting on the fluid, e.g. the cross section of the tear-shaped object in Figure 20.2, which depends on the square of the linear dimension. However, the mass of the body depends on the cube of the linear dimension. If the linear dimension were doubled — i.e. the body is doubled in length, width and breadth — then the force would increase by a factor of four while the mass would increase by a factor of eight; the greater the size of the body the less is the force per unit mass — or its deceleration. For very large bodies, such as planets, normal viscous resistance would be insignificant, although it is the type of resistance that operates on artificial Earth satellites.

20.3 Mass-Dependent Resistance

We now consider resistance where it is the mass of the moving object that is the major influence. Let us suppose that the body is not moving through a conventional fluid but through a region consisting of a large number of more-or-less uniformly spaced solid objects that are so far apart they do not interact with each other. In this case no actual fluid medium exists and any forces of solid objects on each other, due to gravity, depend on their distances apart and not on their motion. If a body moves through this sea of solid objects, which are initially stationary, without touching any of them then, because of the mass of the body, after it has passed through a region the objects are moving, which means that they have energy of motion that they did not have originally. Some of this energy may come from the rearrangement of the solid objects; a system of gravitationally interacting solid objects possesses energy, *gravitational potential energy*, which depends on the positions of the objects. The remainder of the energy of motion of the objects comes from the body that passed through it, which loses energy by slowing down. This slowdown is due to a resisting force that is caused by its own mass acting on the mass of the medium around it. If the medium is a normal fluid then, of course, viscosity must also be present but this new force has nothing to do with viscosity. For the mass-dependent force, theory shows that, in many situations, the rate of deceleration of the body is proportional to its own mass. The density of the medium also plays a role in this resistance mechanism in that the greater is its density the greater is the energy it gains and hence the greater is the resistance. Another way that the density of the medium has an effect, especially if it is very dense, is by the passage of the body causing an uneven distribution of the medium and by the mass clumps so formed having a direct effect on the body due to their gravitational attractions.

For massive bodies, such as planets, the mass-induced resistance forces are dominant and normal viscosity can be ignored. The conceptual model used to explain this kind of resistance, that of having well-separated solid bodies, can also be used to computationally simulate the effect of a resisting medium, which is modelled by a distribution of point masses orbiting the star.

20.4 The Form of the Resisting Medium

We now consider the form of the resisting medium. The region of gas orbiting a star is usually described as a disk but it is dynamically impossible for the gas to take the form of a true disk of uniform thickness. The forces on the gas are not just those of gravity but also the pressure of the gas itself and for this reason the thickness of the disk must increase with distance from the star. Figure 20.3 show the plan view and cross section of a disk, consisting of point masses, used to model a resisting medium, based on a theoretically-derived profile.

In the simulations of orbital evolution, a model planet orbits within the resisting medium with gravitational forces acting between the planet and the particles. The most common outcome is that the orbit decays (becomes smaller) and also rounds off (reduces

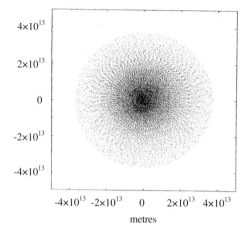

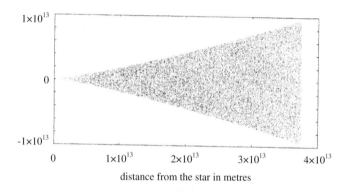

Figure 20.3 The distribution of the particles representing the medium, in the mean plane of the disk (above) and in cross section (below).

eccentricity and becomes more circular). A factor that has to be included in these calculations is the finite lifetime of the disk. The disk material is affected by the radiation from the central star and also the stellar wind, the emission of charged particles by the star. When stars are young they are very active so the disks may disperse quite quickly on an astronomical timescale. Observations of disks around young stars suggest that their lifetimes are from one to a few million years. In the orbital-evolution calculations this characteristic of the disk is introduced by giving the disk a *half-life*. For example, if the half-life is one million years then the density at all points of the disk is continuously reduced so that after that period the density has decreased by a factor of two. After a further million years it is halved again, to one quarter of its original density. The influence of the disk is thus gradually reduced until it becomes insignificant.

20.5 Simulations of Orbital Evolution

We now describe one of these calculations. The medium had total mass 50 M_J (about 0.05 solar mass) with a density that fell off with increasing distance from the star. The half-life of the medium was 1.7 million years. The planet started with a semi-major axis of 2,500 astronomical units and eccentricity 0.9; after 3.7 million years the orbit became almost perfectly circular with a semi-major axis of 5.3 astronomical units, very close to that of Jupiter. It is not claimed that this scenario applied to the actual Jupiter — it just illustrates the way that extreme orbits can evolve to those now observed. Figure 20.4 shows the evolution of semi-major axis and eccentricity with time. Because the range of the semi-major axis is so large the vertical plot is done in logarithmic form. This means that changes by a factor of 10, e.g. from 1,000 to 100, from 100 to 10, or from 10 au to 1 astronomical unit, all correspond to the same distance along the axis.

Some exoplanets are very close to stars, at one tenth of the distance of Mercury from the Sun or even less, while indications from dusty disks and from direct imaging show planets at distances up to 100 astronomical units or so. The decay of an orbit depends on the mass of the planet and also on the total mass of the medium, its distribution and its duration. The higher the density of the resisting medium the greater will be the rate of orbital decay, and the longer the duration of the medium before it dissipates, the greater will be the total decay. The calculation giving Figure 20.4 took the medium mass as 50 M_J but some capture-theory simulations give a retained medium with greater mass, which would give a greater rate of decay of the orbit. The same calculation took the half-life of the medium as 1.7 million years but observation shows that some disks decay much

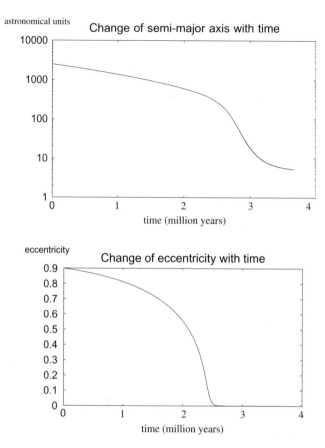

Figure 20.4   Change of semi-major axis and eccentricity with time in a resisting medium.

more rapidly than that, which would reduce the total time for decay to take place. It is clear that various combinations of the medium density and the medium duration could give planets either approaching a star closely or being left stranded at some distance from the star.

Several decay simulations give final semi-major axes less than 0.1 au, as is observed for some exoplanets, and the problem arises of whether planets could plunge into the star. This might happen but there is an interesting mechanism that can prevent it from happening, a mechanism that actually plays a role in a different context within the present Solar System.

The motion of a planet towards the star is in the form of a shallow spiral. When it is close to the star both the planet and star become distorted by tidal forces. Here we are concerned with the star's distortion. The mechanism depends on two periods — one that of the planet in its orbit and the other that of the star's spin, where the latter is smaller than the former. For example, if we take a star of solar mass with a planet in a circular orbit of radius 0.05 astronomical units then the orbit has a period of about four days. We now take the spin period of the star as three days — about nine times less than that of the present Sun but that is not unreasonable for a young star. The gravitational effect of the planet on the star is to produce a tidal bulge on it that will tend to face the planet. However, the faster spin of the star drags the bulge in a forward direction as is illustrated in Figure 20.5.

The force on the planet due to the closely-spherical bulk of the star points towards the star's centre and neither adds to, nor subtracts from, the rotational motion and energy of the planet's orbit. The force due to the bulge is mostly towards the centre but has a small component in the direction of the planet's motion. This gives an extra push to the planet in the direction it is already moving, thus adding energy to the motion of the planet, which tends to move it outwards. This outward-acting effect can just balance the inward-acting effect due to the resisting medium so the planet's orbit is stabilised and it will go no closer to the star. This situation is quite stable. If the planet were to move inwards then the tidal effect would become stronger and push it out again. Conversely, if it happened to move outwards then the resistance force of the medium would dominate and move it in again. The planet would be saved from destruction!

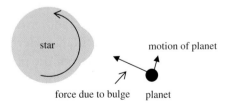

Figure 20.5 Forces on an orbiting planet due to the tidal bulge on a rapidly spinning star.

In the solar-system context, the same kind of outward force acts on the Moon that, as is well known, creates tides on the Earth, large ones in the sea and less obvious smaller ones on land. The spin period of the Earth, one day, is less than the orbital period of the Moon, 29 days, so the tides are dragged forward as shown in Figure 20.5. In this case the push given to the Moon tending to move it outwards is not opposed by a resistance force in the opposite direction and the Moon is receding from the Earth at a rate of 3.8 centimetres per year. The rate was obviously greater in the past when the Moon was closer to the Earth and raised bigger tides on it.

20.6 Eccentric Orbits

A feature of some exoplanet orbits that was somewhat unexpected is that they are highly eccentric with eccentricities up to about 0.9. The standard solar-nebula model suggests that the initial orbits of planets should be near-circular. Although the capture-theory model gives *initial* orbits that are very eccentric the results in Figure 20.4 suggest that, by the time they have decayed to the values of observed semi-major axes, they should have rounded off. To understand how decayed, but eccentric, orbits can occur we need to consider the way in which the forces due to the medium affect the orbit.

The assumption that gave Figure 20.4 is that the resisting medium is in free orbit around the star — that is to say that all parts of it are in circular orbits in similar motion to that of a planet at the same distance. For a planet moving in this resisting medium in a *circular* orbit the local medium speed would be similar to that of the planet, a little faster inwards and a little slower outwards. However, for a planet in an *elliptical* orbit there are substantial relative speeds of the planet with respect to the medium, particularly at periastron[a] and apastron.[a] The general rule that operates with any kind of resistance mechanism is that the force direction always opposes the motion of the planet relative to the medium and the force magnitude is greater for greater relative speed.

---

[a]The terms periastron and apastron are the nearest and furthest points of an orbit from a star (Corresponding to perihelion and aphelion when the star is the Sun).

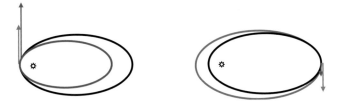

Figure 20.6 The modification of an elliptical orbit in a freely-rotating medium showing the effect of forces at periastron and apastron. At periastron the effect is to modify the orbit towards the green form. At apastron the effect is to modify the orbit towards the blue form.

Figure 20.6 shows the forces at periastron and apastron acting on a protoplanet in an elliptic orbit (shown black) in a freely-rotating medium. At both periastron and apastron the orbital speed is marked in red and the medium speed in turquoise. At periastron the effect of the different speeds is to slow down the planet and modify its orbit to the green form, which has a smaller semi-major axis but the same periastron. At apastron the speed of the medium is greater than that of the planet, which adds speed to the planet and modifies its orbit to the blue form, which increases the semi-major axis but has the same apastron. At both extremes the effect is to round off the orbit but there are opposite effects on the size of the orbit. If, as is usual, the density of the medium is larger closer in, then the periastron effect will be the stronger and the orbit will round off and decay. Of course, there are resistance forces on the planet at all points on the orbit but the essential features of the orbital modification can be understood just by considering the effects at periastron and apastron.

It has already been mentioned that new stars go through a very active stage where they are more luminous and have stronger stellar winds than when they are on the main sequence. For example, it has been estimated that the early Sun could have been 60 times as luminous as now and could have had solar winds between ten thousand and one hundred thousand times as strong as at present. The orbital speed of a body in a circular orbit of a particular radius depends on the strength of the gravitational field at the position of the body; the larger the field is, the greater the speed is. A strong early stellar wind would have applied an outward force on the resisting medium that opposed the gravitational attraction of the star. Extremely strong stellar winds, believed to be present in young stars in the so-called T-Tauri stage of their development, could overwhelm the gravitational attraction of the star and drive the resisting medium outwards. Here we are just going to consider the situation where the stellar wind neutralises some part of the stellar attraction so that the net effect on the medium is as though the star had a reduced mass. In this case the medium rotates more slowly than if the stellar wind were absent. However, no matter how strong the stellar wind, the planet's orbit would be virtually unaffected — the wind force would be insignificant relative to the gravitational force of the star. In Figure 20.7 there is shown the situation where the medium has been heavily slowed down and we show the speeds of the planet and the medium at periastron and apastron. The colour coding of the various speeds at periastron and apastron is as for Figure 20.6; planet speeds are red and medium speeds are turquoise.

At periastron the effect is as previously described — the planet is slowed down, the orbit decays and becomes less eccentric. However, the position at apastron is very different from what it was previously. Now, at apastron, because the medium has been greatly slowed down, the planet is moving faster than the medium, unlike the situation in Figure 20.6. Consequently, the planet is again *slowed down*, the orbit decays and the eccentricity *increases*. The decay is a consistent feature at both extremes of the orbit, but the changes in eccentricity oppose each other. In the most common situation the density is higher and the resistance force stronger at periastron so the effect

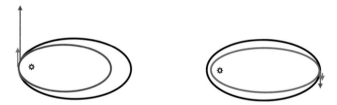

Figure 20.7   The effect on the orbit of a planet moving in a resisting medium heavily influenced by a stellar wind.

there dominates and the orbit is rounded off, albeit more slowly than with a freely-rotating medium. However, in some capture-theory simulations the captured material forming the resisting medium takes up a doughnut-like shape, as is clearly shown in Figure 19.3. This means that when the orbit reaches a certain stage in its development the medium density is *higher* at apastron that at periastron and it is the effect *there* that dominates, so that the eccentricity *increases*. This outcome has been reproduced in numerical simulations and some illustrative results are given in Figure 20.8. The mass of the

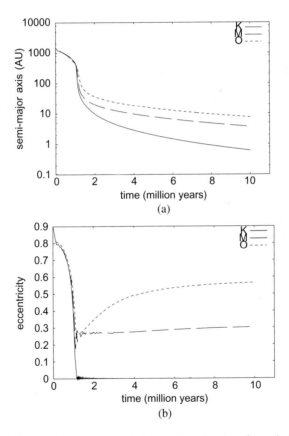

Figure 20.8   Three simulations of orbital evolution showing the variation with time of (a) the semi-major axis in astronomical units and (b) the eccentricity. Two simulations give eccentric orbits.

resisting medium was taken as 50 M_J but a larger mass, which observations would comfortably allow, could considerably shorten the timescales.

In Figure 20.8 simulation K gives a circular final orbit but the other two simulations give ellipses, one with eccentricity nearly 0.6. By assuming a strong 'doughnut' form to the medium even higher eccentricities are possible.

20.7 Commensurate Orbits

A final aspect of orbital evolution in a resisting medium applies to the planetary orbits in the Solar System. When there are several planetary orbits evolving simultaneously then, in the process of the rounding-off and decay, the protoplanets are influenced not only by the Sun and the resisting medium but also, in some circumstances, by each other. That this might be so is suggested by the fact that the ratios of the orbital periods of pairs of major planets are very close to the ratio of small integers. For example,

$$\frac{\text{Orbital period of Saturn}}{\text{Orbital period of Jupiter}} = \frac{29.46\,\text{years}}{11.86\,\text{years}} = 2.48 \approx \frac{5}{2}$$

and

$$\frac{\text{Orbital period of Neptune}}{\text{Orbital period of Uranus}} = \frac{164.8\,\text{years}}{84.02\,\text{years}} = 1.96 \approx \frac{2}{1}$$

There is a mechanism that operates when the orbits of pairs of planets have become circular and are decaying at different rates. When the orbits become *commensurate*, that is, the ratio of their periods equals the ratio of two small integers, then an energy exchange takes place between them. This works in such a way that, although the two planetary orbits continue to decay, they do so coupled together so that the ratio of their orbital periods remains constant. The effect is a subtle one; while a qualitative theoretical explanation is possible, the mechanism is best explored by computation. The ratio of periods for

Uranus and Saturn, 2.85, is not commensurate. Given that the resisting medium evaporates away then it is possible that it simply did not last long enough for a Uranus:Saturn commensurability to become established — although, given time, the ratio may have become either 2.5 or 3.0.

20.8 The Survival of Planetary Systems

In Chapter 19, when dealing with the proportion of stars with planetary companions, the point was raised that the embedded cluster environment was not only conducive to planet formation but also to the disruption of planetary systems. If a newly formed planet was close to apastron in an extended orbit then the near passage of a star could pull the planet away from its parent star and convert it into a free-floating planet. The probability of a star losing some or all of its newly acquired planets depends on the process of orbital evolution. If a planetary orbit stayed in an extended state permanently then the probability that it would be retained within the environment of an embedded cluster would be close to zero. However, once a planet has decayed into a close orbit, similar to those of the solar-system planets or observed exoplanets, then the probability that it would ever be lost is close to zero. It is all a question of timing. If the planet's attachment to the star can survive for the whole time its orbit is decaying then it will become a permanent member of that star's planetary system.

Computational simulations, covering a range of situations, show that between one third and two thirds of planets would be removed from their parent stars. However, about 90 per cent of stars retain *at least* one planet so that the proportion of stars with exoplanets is little affected by the ravages of the embedded cluster. The combination of the probability of forming planetary systems by the capture-theory mechanism together with the probability of disrupting them gives an expected proportion of stars with exoplanets that comfortably satisfies, and even greatly exceeds, the minimum 0.07 observational estimate.

Chapter 21

Now Satellites Form

The known planets consist of the exoplanets that have been detected and imaged, plus the planets of the Solar System. The range of semi-major axes and eccentricities for exoplanets encompasses the ranges found in the Solar System, from 0.39 to 30.1 astronomical units and 0.0068 to 0.206, respectively. Only in the range of planetary masses is the Solar System unlike what has been found for exoplanets, but this may be due to observational limitations; planets with masses as low as those of the terrestrial planets are outside the range of detection at present.

Another solar-system feature that has no counterpart in exoplanet observations is that the major solar-system planets have extensive satellite systems — and even two of the terrestrial planets have satellites, although these have unusual characteristics. Do satellites always occur as a concomitant of planet formation? We cannot answer that question with complete confidence. However, if we make the reasonable assumption that solar-system planets and exoplanets are formed similarly, then, since *all* the major solar-system planets have satellites, we might deduce that the answer to the question is that they do.

21.1 An Outline of the Solar Nebula Theory

The mechanism to be proposed here for the formation of satellites is similar to the mechanism for forming planets by the 'standard model' — the Solar Nebula Theory. Before giving a description of

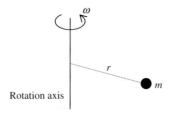

Figure 21.1   A small mass, m, rotating around an axis at distance, *r*, with angular speed, *ω*.

that model we first describe in general terms an important physical quantity, related to the rotation of bodies, called *angular momentum*, the concept that was the basis of the rotating-chair experiment described in Section 12.5.

In Figure 21.1 the point mass *m* rotates around a spin axis at distance *r* with an angular speed *ω* (e.g. radians[a] per second). The angular momentum associated with the body's rotation is given by:

$$\text{angular momentum} = \text{mass } (m) \times \text{distance } (r) \text{ squared} \times \text{angular speed } (\omega)$$

An extensive body can be considered as a sum of a large number of small bodies and then the angular momentum for the whole body is just the sum of the angular momenta for all the small bodies added together. What makes angular momentum so important is that it is one of the *conserved quantities of physics*. Another conserved quantity is energy — it cannot be destroyed but can be converted from one form to another — mechanical to heat, electrical to mechanical, etc. Einstein's contribution was to add mass to the possible forms in which energy can occur. Conservation of angular momentum means that, for any system involving rotating masses, which is isolated in the sense that it is not acted on by external forces, the total angular momentum must remain constant.

The 'standard model' of planet formation begins with a rotating dusty gaseous sphere collapsing under the influence of gravity

---

[a] The radian is the angle subtended at the centre of a circle by an arc on its circumference of length equal to the radius. It is about 57.3°.

(Figure 21.2(a)). As the gaseous body collapses its material moves closer to the rotation axis. Since angular momentum remains constant, to compensate for material getting closer to the axis (r reducing), the spin rate (ω) must increase. This is a well known phenomenon in the realm of exhibition ice skating. As a skater, gracefully pirouetting on her skates with arms outstretched draws in her arms, then since some mass is closer to her rotation axis the gentle pirouette becomes a rapid spin. For the rotating mass of gas, as it spins faster, so it flattens along its spin axis (Figure 21.2(b)). Eventually, as the nebula collapses, material at the boundary of the spinning mass rotates so quickly that it is in orbit around the central mass and thereafter a disk of material is left behind by the collapsing core (Figure 21.2(c)). The central core eventually becomes the star and planets are produced from the material of the disk.

This theory, first put forward by Pierre-Simon Laplace (Figure 21.3), was described in great detail in his book *Exposition du Système du Monde* in 1796 and this model for Sun and planet formation was widely accepted for more than 100 years. Eventually it became discredited, mainly due to an objection based on angular momentum. The Sun contains 99.86% of the total solar-system mass, the rest being mainly in the planets. However, the Sun spins slowly and the planets have large orbital radii so the Sun, for all its dominance in terms of mass, contains only 0.5% of the angular momentum of the system. To summarise — the Sun has 700 times the mass of all the planets combined but the planets together have 200 times as

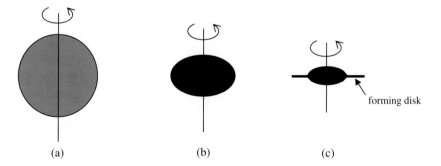

Figure 21.2 Stages in the collapse of a spinning dusty gaseous sphere.

Figure 21.3   Pierre-Simon Laplace (1749–1827).

much angular momentum as the Sun. No way could be found to explain this partitioning of mass and angular momentum beginning with any reasonable starting configuration of the gas sphere. In its recent rebirth the Solar Nebula Theory addresses this problem, either by calling on the action of magnetic fields to transfer angular momentum from inner to outer material or by invoking rather complex mechanical processes. It is fair to say that the problem has been side-stepped rather than solved.

Detailed and well-founded theoretical work on planet formation from disk material has been developed by Solar-Nebula theorists. The steps in this process are as follows:

Step 1

The dust settles down under gravity to form a thin carpet in the mean plane of the disk. The dust consists of tiny particles, one micron (one millionth of a metre) or less in diameter, which makes the process of settling very slow. However, particles can stick together to make larger aggregations that will settle more quickly.

Step 2

The carpet of dust becomes gravitationally unstable and forms clumps (Figure 8.5). These clumps form solid bodies, called *planetesimals*, the size of which depends on the local conditions in the disk but they have dimensions in the kilometre to 100 kilometre range.

Step 3

Planetesimals aggregate to form larger bodies that in the inner region of the Solar System become terrestrial planets and in the outer region form the silicate-iron cores of the major planets. The theory for this aggregation process is well developed and was first described by a Russian planetary scientist, Victor Safronov (Figure 21.4).

Step 4

The cores of the major planets acquire atmospheres by capturing gas present in the nebula disk.

The process of forming the dust disk (step 1) presents some theoretical difficulties but not severe enough to completely discredit the formation process. The break-up of the dust disk (step 2) is

Figure 21.4 Victor Safronov (1919–1999).

theoretically well founded and presents no obvious problems. Safronov's theory (step 3) is sound and planetesimals would gather together as he suggests. Finally, the accumulation of gas (step 4) to form the bulk of the major planets is straightforward and would happen on a fairly short timescale. So, is all well for planet formation? Unfortunately not — the problem is one of timescale. In applying Safronov's theory to any reasonable original disk it would take one million years to produce the Earth, 200 million years to produce Jupiter and a staggering 10,000 million years to produce Neptune. Since disks around young stars are observed to have lifetimes up to a few million years at most, then the difficulty with the theory is obvious. Attempts have been made to overcome the problem. Conditions in the disk have been postulated that speed up the process of accumulating planetesimals, although the conditions are somewhat contrived and proposed, not because there is some reason to expect them, but just for the purpose of trying to solve this particular problem. The basic difficulty is one of forming planets at large distances from the Sun, and no conceivable conditions, however outlandish, can give reasonable timescales for the formation of Uranus and Neptune.

The main approach to trying to solve this problem is the theory that the outermost planets were formed much closer to the Sun but then migrated outwards due to the joint action of Jupiter and a resisting medium. We know that a resisting medium can take energy of motion *away from* a planet and so cause its orbit to decay. The process of *adding* energy to a planet to send it from, say, the region of Jupiter to that of Neptune is far trickier, if actually possible. The proposed mechanism is that Jupiter moves inwards losing angular momentum that is given to outwardly moving resisting-medium material in the form of a wave, much as a boat moving through water produces an outwards wave from its bows. This wave then impinges on a planet further out, transfers some of its angular momentum to the planet and hence propels it outwards. It is argued that, since Jupiter is so massive compared to the other planets, a small inward movement by Jupiter can provide all the extra energy and angular momentum necessary. A weakness in the mechanism seems to be that the planet on which the spiral wave impinges is a very small target and will thus take up very little of the angular momentum in the wave.

21.2 The Formation of Satellites

Before discussing the origin of satellites, we first highlight an important difference between the relationship of satellites to planets and that of planets to the Sun. In what follows we refer to the larger central body as the *primary body* and the one in orbit around it as the *secondary body*. For the two kinds of system — satellite:planet and planet:Sun — we take two ratios. The first is the ratio of the angular momentum *per unit mass* of the orbiting secondary body to the angular momentum *per unit mass* of material at the equator of the spinning primary body. The second is the ratio of the orbital radius of the secondary to the radius of the primary. Table 21.1 shows these ratios for various primary:secondary pairs.

Both ratios S and R show the difference between the two kinds of system; to summarise, the satellite:planet systems are far more

Table 21.1 The ratios.

$$S = \frac{\text{Angular momentum per unit mass of secondary in orbit}}{\text{Angular momentum per unit mass of primary at equator}}$$

$$\text{and} \quad R = \frac{\text{Radius of orbit of secondary}}{\text{Radius of primary}}$$

Primary	Secondary	Ratio S	Ratio R
Sun	Jupiter	7,800	1,120
Sun	Neptune	18,700	6,458
Jupiter	Io	8	5.9
Jupiter	Callisto	17	26.3
Saturn	Titan	11	20.3
Uranus	Oberon	21	22.8

compact. When Galileo saw the Galilean satellites around Jupiter it consolidated his belief in the Copernican model of the Solar System but, despite the superficial resemblance, the two systems are clearly quite different in important characteristics.

Figures 19.2, 19.3 and 19.4 show SPH simulations of planet formation according to the Capture Theory. The planets are shown having just formed but with no indication of how they subsequently evolve. The development of a planetary condensation, taken from an SPH calculation, is illustrated in Figure 21.5. It shows the path of the planet and its collapse over a 1,000 year period at intervals of 100 years. There is considerable collapse in this period and, at the end, slightly more than one half of the material forms a condensed core while the remainder forms a surrounding disk. The scale of the condensation at this stage is that the central core has a radius which is of the order of one-half of an astronomical unit and the disk has a

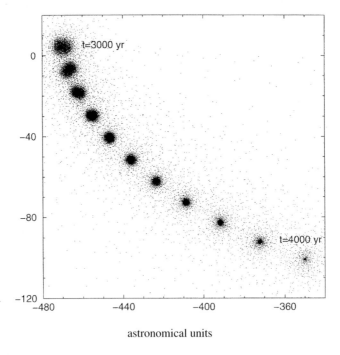

astronomical units

Figure 21.5   The collapse of a protoplanet.

radius of about 4 astronomical units. The rate of collapse of the core is high at the end of the simulation — indeed it is following a free-fall pattern — and it will not take long to fall to near-planetary dimensions.

The disk round the planet is much denser than that in the solar-nebula model of planet formation. The areal density of the solar-nebula disk is about 10 kilograms per square metre while that of the disk around the planetary core is 4,000 kilograms per square metre. This illustrates the general point that it is in the concentration of matter that the Capture Theory contrasts so starkly with the Solar Nebula Theory. In forming planets the whole mass of a compressed region or of a protostar is pulled into a compact filament within which the Jeans mechanism for gravitational instability can operate. By contrast, in the 'standard model' the material is thinly distributed within a large volume and must come together by slow processes of aggregation.

The conditions in the disk are suitable for the first three steps that were described for forming planets in a nebula to occur, except that now satellites are formed for which the fourth step, acquiring a gaseous atmosphere, does not normally happen. The computational outcome of what these steps give will now be described. The planet is taken with mass 1 M_J with a disk of the same mass. The mass per unit area of the disk is a maximum towards the centre and falls off such that it is halved for each distance of one half of an astronomical unit from the planet.

Step 1

The planetary collapse generates a high temperature and the resultant radiation pressure tends to counteract gravitational forces. Another factor is the temperature of the disk material due to the luminosity of the planet, in relation to the escape speed at different distances. Calculation of the balance of these forces shows that material further from the planet than 20 million kilometres will be lost on a timescale less than that required for a dust layer to form. The dust within this distance from the planet will settle on a timescale of about 9,000 years and will have a total mass of 5–6 lunar masses, approximately the combined masses of the Galilean satellites.

Step 2

The dust disk rapidly breaks up through gravitational instability to form *satellitesimals*, the equivalent in this context to planetesimals. Close to the planet the satellitesimals form in a few days and they have masses of 10^9 kilograms (one million tonnes). At a distance of 20 million kilometres the time rises to about two years and the masses of the satellitesimals rise to about 10^{17} kilograms (about one millionth of the mass of the Moon).

Step 3

The calculated formation times of satellites from satellitesimals by the Safronov process varies from a few thousand years close in to about a million years at 20 million kilometres. However, the latter time is certainly an overestimate since it assumes that satellite aggregation took place from start to end at a distance of 20 million kilometres, which would not be true. The satellitesimals form within a resisting medium, and their orbits and those of the growing satellites would be in a state of constant decay. If the model we are considering is that of forming the Galilean satellites then the outermost one, Callisto, must eventually end up at a distance of 1.88 million kilometres. The scale of decay of orbits required for satellites is less than that for planets but then so is the time for decay processes to operate.

We have noted that the commensurate orbits of the pairs of planets Jupiter:Saturn and Uranus:Neptune are explained by a coupling of their decays in a resisting medium. The Galilean satellites Io, Europa and Ganymede are triply commensurate, with orbital periods closely in the ratio 1:2:4 and with an exact relationship linking the three periods. This commensurability, and that of pairs of Saturnian satellites, can be explained by the same mode of coupling during decay.

The above has described a mechanism for the formation of regular satellites that orbit their planets in near-circular orbits in the equatorial plane. Other satellites are probably captured bodies but the process of capture requires some explanation. If a planet and a satellite come together from a great distance apart then, in general, they will approach each other and then move apart to the same large distance. Capture will not take place. To have capture, a process that can remove energy of motion from the two-body system is needed. This could come from a collision, where energy of motion is converted into the energy to heat the bodies and break them up. A near approach giving large tidal forces would also turn some energy into heat and might give capture, although this process would be much less effective than a collision. Another and very effective way of removing energy of motion from the two bodies is if a third body is in the vicinity, which takes up extra energy and removes it from the two bodies.

An example of non-regular satellites is the two outer groups of satellites of Jupiter (Table 16.1). Those in direct orbits have semi-major axes between 11 and 12 million kilometres and those in retrograde orbits semi-major axes between 21 and 24 million kilometres. However, because of their orbital eccentricities the perijove (closest distances to Jupiter) of some of the outer retrograde satellites are closer to Jupiter than the apojove (furthest distances from Jupiter) of some of the inner group. For example the perijove of Ananke is 12.6 million kilometres while the apojove of Elara is 13.7 million kilometres. The origin of these two groups of satellites is probably due to a collision between two asteroids in the vicinity of Jupiter (Figure 21.6). The asteroids, approaching as shown, are shattered and, with reduced energy of motion, are captured — group A in a direct sense and group B in a retrograde sense. The orbits are perturbed by other bodies — the Sun, other planets and other satellites — so they will have drifted away from their original

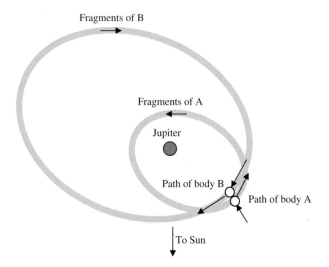

Figure 21.6   A schematic representation of a collision of two bodies near Jupiter giving rise to the two outer families of satellites.

orbits while still retaining the characteristic that in their distances from Jupiter they overlap.

Speculative collision scenarios can be suggested to explain some other non-regular satellites — sometimes more than one possibility for a particular satellite. However, some exceptional satellites obviously require detailed explanations, notably the Moon and Triton, the large retrograde satellite of Neptune. We return to these later.

Chapter 22

What Can Be Learnt from Meteorites?

22.1 The Origin of Asteroids

In Chapter 17 meteorites were identified, for the most part as samples of asteroids but occasionally coming from other bodies, such as Mars. There are a few meteorites that are suspected of being of lunar origin, chunks knocked off the Moon by projectiles at some time. All meteorites are samples of solar-system bodies, obtained at small expense, and as such offer valuable information about those bodies that would be impossible, or at least difficult and expensive, to obtain in any other way.

In terms of their significance as members of the Solar System, asteroids can be explained in two different ways. One interpretation is that they are the building blocks from which planets formed. The sizes of most asteroids fall within the size range of planetesimals (Section 21.1) and they might be thought of as residual planetesimals after the planets formed. The total mass of all known asteroids is equivalent to that of a large satellite so, on this interpretation, they could be regarded as just the crumbs left over from the planet-making process. Another possibility, suggested long ago, is that they are the debris from the break-up of a planet, although, at the time the suggestion was made, there was no believable scenario for a planet spontaneously to disintegrate.

Despite the difficulty of disrupting a planet the idea had credence because of the nature of meteorites. Meteorites are generally classified as irons, stones and stony-irons, and having material separated in this way is just what happens in a planet. By whatever means a planet formed it would begin as a hot, molten body and would inevitably settle into layers according to the density of its constituents. The densest material, iron, would sink to the centre and form a core. This would be surrounded by denser silicates to form the mantle and finally, for a terrestrial planet, the least dense silicates would float like a scum to the surface and constitute the planetary crust. A major planet would finally acquire the gaseous component that, as for Jupiter and Saturn, constitutes the majority of its mass. The disintegration of such a body would give asteroids that were mostly irons or stones with some near-boundary material giving stony-irons such as pallasites (Section 17.3.3). These types of material explain what is observed in meteorites, the fragments from asteroids.

The evidence that meteorites came from differentiated bodies is so strong that it demands an explanation. Planetesimals formed as proposed by the Solar Nebula Theory were likely to have been cold intimate mixtures of iron and stone particles and would be unsuitable sources to explain the composition of meteorites. To resolve this difficulty the idea of *parent bodies*, intermediate-sized bodies formed by collections of planetesimals, was advanced. When planetesimals aggregated to form a large body then the energy of motion of the in-falling bodies was transformed into heat. Later incoming material arrived with greater speed, because it was attracted by the growing body, so that the temperature of deposited material increased as the body size increased. For silicate materials a body of radius about 1,100 kilometres is required before melting begins, and for substantial melting within the body something approaching the size of the Moon is required. A body formed in this way would be massive enough both to generate the temperature for melting and also to provide the gravitational field for the differentiation of material. Collisions of parent bodies, leading to their complete disruption, would then provide smaller, irregularly-shaped asteroids, and perhaps potential meteorites with the required differentiated compositions, and subsequent collisions of asteroids would provide further meteorites. By analysing the compositions of meteorites it was concluded that they could all be explained as material from about 20 parent bodies. The parent bodies gave another benefit because they were massive enough to give internal pressures that could explain some of the minerals found in meteorites that can only form at high pressure.

The idea of building up parent bodies and then breaking them up again was not very acceptable to some workers in the field who would have much preferred to make do with planetesimals. They argued that if in some way planetesimals could have been molten then differentiation would have taken place, albeit more slowly than in a larger body. As for the formation of high pressure minerals, such as diamond, the production of high pressures could happen in the form of a shock when the asteroids collided to give meteorites. If only the planetesimals could have been molten! Help was at hand — radioactivity came to the rescue to provide the heat energy required.

22.2 Aspects of Radioactivity

Figure 7.2 shows a representation of the nucleus of deuterium, with one proton and one neutron. Deuterium is not really a distinctive type of atom in a chemical sense — it is an isotope of hydrogen and chemically it *is* hydrogen. If a chemical compound contains hydrogen then any or all of the hydrogen atoms it contains can be replaced with deuterium and it is still the same chemical compound. Water is H_2O but HDO, with one hydrogen atom replaced by deuterium, is also water as is D_2O (heavy water) where both hydrogen atoms have been replaced. If you were given HDO or D_2O to drink you could do so quite happily and not realise that it wasn't ordinary water. More to the point — you drink HDO every day! A proportion of all the hydrogen on Earth, and elsewhere for that matter, is deuterium; in the case of the Earth, 16 of every 100,000 hydrogen atoms are deuterium.

Deuterium is stable and will maintain its identity indefinitely, as will hydrogen. However, a hydrogen isotope called tritium (symbol T) is unstable. Its nucleus contains one proton (that makes it chemically hydrogen) plus two neutrons. The three isotopes of hydrogen are illustrated in Figure 22.1.

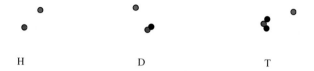

Figure 22.1   The three isotopes of hydrogen (red = proton, black = neutron, blue = electron).

Now we introduce notation that uniquely identifies an isotope — or just the nucleus if the electrons are removed. In this notation hydrogen, deuterium and tritium are:

$$^1_1\text{H}, \, ^2_1\text{D} \text{ and } ^3_1\text{T}.$$

The chemical symbol is the letter. Hydrogen is unique in that these three isotopes have different letter symbols — for example, all the isotopes of carbon are represented by the letter symbol C. The bottom number on the left-hand side of the symbol gives the number of protons in the nucleus that identifies the type of atom so either it or the letter symbol is really redundant. The top number indicates the mass of the isotope, which is the combined number of protons and neutrons. It must be stressed that the chemical nature of an atom is entirely dependent on the number of protons — or perhaps more precisely the number of electrons in the neutral atom, since it is electrons that are involved in chemical reactions.

The statement that tritium is unstable means that it can, and does, spontaneously break down. A tritium atom disintegrates in the following way:

$$^3_1\text{T} \rightarrow \, ^3_2\text{He} + e^- + \overline{\nu}_e \, .$$

We now translate this decay equation. On the left-hand side of the arrow is the tritium we start with. The first quantity on the right-hand side — helium-3 — is a stable isotope of helium, of which the most common stable isotope is ^4_2He with two protons and two neutrons. The way that the decay happens is that one of the neutrons in the tritium nucleus transforms into a proton plus an electron. The proton is retained in the nucleus, to make it helium-3, and the electron, symbol e⁻, shoots out of the nucleus as a β-particle. The last particle on the right-hand side is a neutrino — or, to be precise, an *electron antineutrino* — but we need not worry about niceties of notation where neutrinos are concerned. Although tritium is unstable, individual atoms may persist as tritium for some time. One cannot predict when any particular atom will decay but what one *can* predict is that, taking a very large number of such atoms, one half of them will decay in a period of 12.3 years. This period of time is called the *half-life* of tritium. After 12.3 years only half the original tritium will remain and after 24.6 years only one quarter of the original atoms will remain. Each period of 12.3 years thereafter sees a further reduction in the number of remaining tritium atoms by a factor of two. Because of this short half-life any tritium that exists at any time virtually disappears after a few hundred years, so it is not a naturally-occurring isotope. It has medical applications and is potentially useful material for producing power by fusion but the only way of obtaining tritium is by producing it within a nuclear reactor.

22.3 Melting of Planetesimals

The description of isotopes provides not only an explanation of how small asteroids could have melted but also provides a background to explain many other aspects of meteorite composition. To explain melting, the element of interest is magnesium, with 12 protons in its nucleus. The three stable isotopes of magnesium are:

$$^{24}_{12}\text{Mg}, \ ^{25}_{12}\text{Mg and} \ ^{26}_{12}\text{Mg}.$$

with 12, 13 and 14 neutrons in their nuclei, respectively. In normal magnesium, such as might be extracted from a mineral or sea water, the proportions of these three types of magnesium are:

$$^{24}_{12}\text{Mg} : \ ^{25}_{12}\text{Mg} : \ ^{26}_{12}\text{Mg} \ = \ 0.790 : 0.100 : 0.110.$$

In 1976 three American meteoriticists (specialists in the study of meteorites), T. Lee, D. Papanastassiou and G. Wasserberg were studying isotopic compositions in the white CAI inclusions in carbonaceous chondrites (see Section 17.3.1). When they measured magnesium isotopes they found an excess of magnesium-26. Different grains of a particular meteorite specimen have different chemical contents so, from one grain to another, the magnesium content varied. What the investigators found is that while the excess of magnesium-26 varied from one mineral grain to another the excesses *were proportional to the amount of aluminium present in the sample.* The only stable isotope of aluminium is $^{27}_{13}Al$ (13 protons plus 14 neutrons). However, a radioactive isotope, aluminium-26 ($^{26}_{13}Al$), with one less neutron and a long half-life of 720,000 years, disintegrates to give magnesium-26. The interpretation of the observations was as follows. Once the CAI inclusion had solidified nothing was able to escape from it and it contained a certain amount of aluminium. Most of the aluminium was the stable aluminium-27 but a tiny fraction of it was the radioactive aluminium-26. Different amounts of aluminium went into different grains and all the aluminium-26 decayed over a long time period to give magnesium-26. All the grains contained some magnesium so when the magnesium isotopes were examined the excess of magnesium-26 was proportional to the total amount of aluminium in the grain. The important consideration here is that the decay process produces heat. The inferred proportion of the aluminium that was the radioactive isotope is from 10^{-5} down to 10^{-8} for different mineral samples. At the upper end of that range the amount of heat generated would be sufficient to melt asteroids just a few kilometres in diameter. This explanation for the melting of asteroids poses the question of the source of the aluminium-26. It is likely that, however the Solar System formed, it would have been preceded by a supernova event that would have triggered formation of a star-forming cloud. Many isotopes, including some that are radioactive, are produced in a supernova and, as long as the Solar System formed within a few half-lives of aluminium-26 after the supernova, there would have been enough of it around to give the required heating. If aluminium-26 was contained within asteroids when they formed then there is no need to postulate the formation of large parent bodies.

22.4 Details of Meteorite Composition

In Chapter 17 several characteristics of meteorites were described that clearly have something to say about their origin. At some stage silicate materials were not only molten at a very high temperature, which explains very small chondrules, but must even have been in a vaporised state. Many of the characteristics of the minerals in meteorites can be explained by the sequence in which minerals appear, first in liquid and then in solid form, as a silicate vapour cools. It can also be inferred that this vaporisation was produced by some sudden explosive event rather than being due to a long-lasting very-high temperature state of the Solar System. We know this because an examination of chondrules shows that they must have cooled very quickly. When a chondrule formed as a molten drop the minerals within it broke up into various stable components. For example, olivine (Mg_2SiO_4), a very common mineral in the mantle of planets, in liquid form breaks down into the units $2MgO + SiO_2$, where MgO is magnesium oxide and SiO_2 is silicon dioxide, generally called *silica*. However, these units are also the components of many other minerals. When the chondrule is solid, but still very hot, the units have enough energy to move around to produce different combinations of minerals. Given enough time, this chemical gavotte produces the most stable configuration possible and when that state is reached the collection of minerals is said to be *equilibrated*. However, the minerals in chondrules are *unequilibrated*, the reason being that they solidified and then cooled so quickly that the chemical units were trapped in fixed positions in the solid before they could achieve the equilibrated state. Another indication of some explosive event, or events, in the early Solar System is the formation of mesosiderite stony-irons (Section 17.3.3), which required a mixture of fine molten iron and molten stone fragments to come together and quickly solidify.

A more subtle, but important, characteristic of meteorites is their isotopic composition, already mentioned in relation to magnesium.

The characteristic of interest is that of *isotopic anomalies*, meaning the difference between the isotopic compositions of particular elements in a meteorite compared with those on Earth. Isotopic compositions of particular elements are much the same all over the Earth. Small differences can occur due to temperature gradients or chemical reactions but the differences are predictable in the sense that the ratios of the isotopes will vary systematically from one isotope to another. If some solar-system object has an isotopic composition significantly different from that on Earth that cannot be explained by some systematic variation from the Earth values, then this must invite the question of why it is so. For example, in Section 15.3 the high D/H ratio in Venus was explained by the way its atmosphere had evolved, which also explained the aridity of that planet. Some interesting isotopic anomalies found in meteorites are now described, together with some of the explanations that have been made for them.

22.4.1 *Oxygen*

The three stable isotopes of oxygen, $^{16}_{8}O$, $^{17}_{8}O$ and $^{18}_{8}O$ occur in the ratios $0.9527 : 0.0071 : 0.0401$; this is known as SMOW (Standard Mean Ocean Water) and is characteristic of oxygen samples from all terrestrial sources. Some anhydrous (water-free) materials from carbonaceous chondrites have anomalous oxygen isotopic ratios that can be interpreted as the result of adding pure oxygen-16 to normal terrestrial oxygen.

An explanation that has been given for this anomaly is that pure oxygen-16 is produced in stars by a reaction between the common isotope of carbon, carbon-12, and an alpha-particle (helium-4 nucleus). This pure oxygen-16 is then incorporated into grains which drift across space and enter the Solar System. The pure oxygen-16 then diffuses out of the grains and mixes with normal oxygen in some solar-system bodies. If you find this explanation a little far-fetched you will not be alone!

22.4.2 *Carbon and Silicon*

A common mineral in chondrites is silicon carbide, SiC. The stable isotopes of carbon are carbon-12 and carbon-13, $^{12}_{6}C$ and $^{13}_{6}C$, and these occur on Earth in the ratio 89.9:1. However, many carbon samples in silicon carbide from chondrites have much smaller ratios, down to 20:1, so that there is a much greater proportion of the heavier carbon-13. This anomalous carbon, known as *heavy carbon*, has again been explained by grains drifting across space, this time from *carbon stars* — stars that are large, reddish in colour and contain a great deal of carbon. It is suggested that carbon from six or more carbon stars could explain all the observations of carbon anomalies.

There are also silicon isotopic anomalies in SiC grains. The three stable isotopes of silicon are $^{28}_{14}\text{Si}$, $^{29}_{14}\text{Si}$ and $^{30}_{14}\text{Si}$, in the ratios 0.9223: 0.0467:0.0310 in terrestrial silicon. Various silicon carbide grains give large variations from terrestrial silicon. The silicon anomalies are not correlated with the carbon anomalies in any way.

22.4.3 *Nitrogen*

Nitrogen trapped within silicon carbide grains also shows isotopic anomalies. The two stable isotopes of nitrogen are nitrogen-14 and nitrogen-15, $^{14}_{7}\text{N}$ and $^{15}_{7}\text{N}$, with the terrestrial ratio 270:1. In many silicon carbide grains the ratio is as high as 1,400:1, referred to as *light nitrogen*. However, a few samples gave *heavy nitrogen* with a ratio of the two isotopes 50:1.

22.4.4 *Neon*

Neon is a minor component of the Earth's atmosphere. It is chemically inert and found in meteorites, trapped in small cavities from which it is released by heating. The three stable isotopes in normal terrestrial neon, $^{20}_{10}\text{Ne}$, $^{21}_{10}\text{Ne}$ and $^{22}_{10}\text{Ne}$ are in the proportions 0.9051:0.0027: 0.0922. There are many neon samples from meteorites with considerable departures from the terrestrial ratios but the most remarkable samples, referred to as neon-E, are almost pure neon-22.

The most likely source of neon-E is as the decay product of sodium-22, $^{22}_{11}\text{Na}$; there is only one stable isotope of sodium, sodium-23, $^{23}_{11}\text{Na}$. Sodium is contained in various minerals and if when the meteorite was formed a proportion of the sodium in it was sodium-22 then this would decay and the resultant neon-22 would collect in cavities within the meteorite. That is a straightforward story but there is a twist in the tail. The half-life of sodium-22 is 2.6 years, which imposes severe time restraints on any theory of the origin of the Solar System. Sodium-22 has to be created in some energetic event, such as a supernova, which produces many new isotopes including some that are radioactive, then incorporated in a mineral within a meteorite parent body and, finally, that body must become cold within a few half-lives of sodium-22. Remember, neon is extracted from meteorites by heating so if the meteorite was hot for the whole period of existence of the sodium-22 then no neon would be retained. Another explanation offered for neon-E is that in the early Solar System, when solar winds might have been very strong, protons in the solar wind reacted with neon-22, a normal component of neon that was present at the time, to produce sodium-22. The sodium-22, being chemically reactive, then became part of a mineral and subsequently decayed.

The isotopic anomalies described here are not the only ones but they are a selection of important anomalies, the explanations of which are of importance not only to meteoriticists, but also to those interested in the origin and evolution of the Solar System.

Chapter 23

A Little-Bang Theory and the Terrestrial Planets

23.1 The Problem of Terrestrial Planets

The capture-theory process of planetary formation from condensations in a dusty filament was described in Chapter 19. Once a planetary condensation formed, the dust settled towards the centre, giving an iron-silicate core that became hot and molten from the gravitational energy released by its inward fall. The gravitational field within the core then gave the separation of materials by density, with denser iron falling to the centre and a silicate mantle forming around it. The gas was left outside the core, giving a typical major planet. The essential feature of the process is that the total mass of the condensation, which is mostly gas, should be greater than a Jeans critical mass and hence be able to collapse. With this process of planet formation, then how is it that terrestrial planets were formed?

Since the terrestrial planets are those closest to the Sun it has been argued that the temperature at these distances would not allow planets to form with hydrogen-helium atmospheres. That is true for any process, such as that of the Solar Nebula Theory, where a solid core first formed. The Earth could not now retain a tenuous hydrogen atmosphere so there is no way, even if the Earth were immersed in a hydrogen environment, that it would start to acquire one. That does not mean that a planet similar to Jupiter could not *survive* in the vicinity of the Earth — indeed it could. Table 18.1 shows that planets similar to Jupiter exist at distances from Sun-like stars just one tenth of the distance of Mercury from the Sun. If a substantial gaseous planet is formed sufficiently far from a star, in a region where the temperature is so low that a hydrogen atmosphere can be gradually built up starting from nothing, followed by orbital decay to bring it close to the star, then it will survive intact. The Solar Nebula Theory has no problems in this respect — assuming that it can actually form planets in the first place.

The capture-theory model gives planetary condensations above Jeans-mass that initially moved on orbits that took them well away from the star for a few tens of thousands of years. Figure 21.5 shows the rapid collapse of a planetary condensation during a period of 1,000 years and it is evident that when the condensed planet again approached its star — the Sun in the case of the Solar System — it would have been safe from disruption, despite the temperature environment. Given this scenario it seems that the capture-theory process does not give terrestrial-style solid planets and this is an apparent difficulty of the theory that must be confronted. The following discussion will be in the context of explaining the terrestrial planets of the Solar System.

23.2 The Precession of Evolving Orbits

The decay and rounding-off of a planetary orbit in a resisting medium was described in Section 20.5. An orbiting planet in the absence of a resisting medium would repeatedly go through the same points in space relative to the star, i.e. the orbit would not change with time. If we add a resisting medium the orbit decays and, in general, rounds off, although, for a doughnut-shaped medium and a very active star, the orbit may become highly eccentric. In the presence of a resisting medium something else also happens — the orbit will undergo *precession*. This kind of motion, seen in projection looking down on the mean plane of the resisting medium, is illustrated in Figure 23.1 and shows that the major axis of the orbit steadily rotates. To get a full picture of what is happening we must consider the motion in three dimensions. The major axis will be

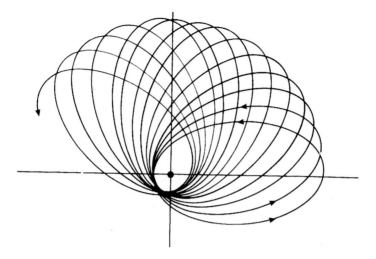

Figure 23.1 The precession of an orbit as seen in plan view.

inclined at a few degrees to the mean plane of the medium (the plane of the figure) and as the orbit undergoes precession this angle remains unchanged.

The precession is due to the gravitational influence of the resisting medium, which is small but not negligible compared with the Sun's gravitational effect. The gravitational force on a planet due to the resisting medium will be offset from the solar direction and hence the total gravitational force, due to the Sun plus the resisting medium, will be slightly offset from the solar direction. This is illustrated in Figure 23.2. It is the component of the net force perpendicular to the plane of the orbit that causes the precession. Computational simulations of this motion show that the period of the precession, the time taken for the major axis to make one complete turn, is a few hundred thousand years, so that there will be a few complete precession periods during the evolution of the orbit to its final state.

The original planetary condensations would have had small, all different, inclinations to the mean plane of the resisting medium. Not all the motions of the material shown as producing planets in Figures 19.2, 19.3 and 19.4 would have been in the same plane. The compressed medium, or the protostar, stretched into the filament

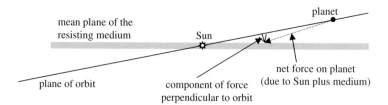

mean plane of the
resisting medium

Sun

planet

net force on planet
(due to Sun plus medium)

plane of orbit

component of force
perpendicular to orbit

Figure 23.2   The force component causing the orbital precession. The deviation of
the net force from the orbital plane has been exaggerated for the sake
of clarity.

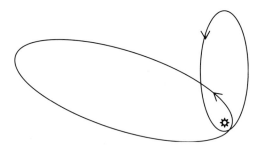

Figure 23.3   Inclined orbits that appear to intersect in projection but do not
actually intersect.

would have had a small component of motion out of the plane of the
figures. Figure 23.3 shows two planetary orbits in projection on the
mean plane of the resisting medium. The orbits appear to intersect
but, if both orbits are inclined to the mean plane, then an exact intersection of the orbits is unlikely. At the points of apparent intersection
the orbits will be separated perpendicular to the plane of the figure.

23.3 A Postulated Planetary Collision

The precession rate depends on the size and shape of the orbit and
its inclination to the mean plane of the medium. If there is *differential* precession of two orbits then the projected angle between the
major axes will change with time and sometimes the orbits *will* intersect in space. This creates the possibility of a collision between the

planets — to get an actual collision the planets will have to be at the same place at the same time, which seems a very demanding condition. Calculations for an initial Solar System containing six major planets in eccentric but non-evolving orbits with differential precession showed that the expectation time for a collision to take place between some pair of planets was about ten million years. Taking into account the actual decay and rounding-off, it is estimated that the probability of a collision for some pair of the six planets (there are 15 possible pairs) during the million years or so of orbital evolution is between 0.1 and 0.2 for the model that was taken — small but by no means negligible. We shall be considering how the assumption that a collision *did* take place could explain many of the observed features of the Solar System.

The six major planets assumed for the original system are the four that now exist plus two others that would have rounded off closer in but collided before they could do so. To consider a scenario of colliding planets we need to know the characteristics of the two planets and how they were moving. The possibilities are many. We know that planets collapse on a short timescale so it is likely that at the time of any collision the planets were close to their present configurations. Given that they would have been major planets they could have had masses anywhere in the range of observed major planets both within and outside the Solar System. Although Jupiter is the largest planet in our system, much more massive exoplanets exist. Later it will be demonstrated that the planetary collision hypothesis explains many features of the Solar System and not just the terrestrial planets. So it is, with an eye to providing these many explanations, that characteristics of the colliding planets (Table 23.1) were chosen. The more massive planet, which we call Bellona, has nearly twice the mass of Jupiter and the other, which we call Enyo,[a] has about 1.2 times the mass of Saturn.

The contact speed of collision was taken as 80 kilometres per second but, since they sped up as they approached each other because of their mutual gravitational attraction, their approach speed when they were far apart was 48.8 kilometres per second. The speed of the Earth

---

[a] Bellona is the Roman Goddess of War and Enyo her Greek counterpart.

Table 23.1   The characteristics of the colliding planets.

	Enyo	Bellona
Mass (Earth units)	116.4	617.6
Radius (kilometres)	6.050×10^4	8.582×10^4
Central density (kilograms per cubic metre)	98,000	162,000
Central temperature (K)	48,000	76,000
Mass of iron (Earth units)	1.875	2.75
Mass of silicate (Earth units)	7.50	11.00
Mass of volatile materials (Earth units)	3.75	5.5

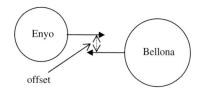

Figure 23.4   The collision of planets with respect to their centre of mass, showing the offset.

in its orbit around the Sun is 30 kilometres per second so, given that the planets were on eccentric orbits, the approach speed of the planets indicates a collision somewhere in the terrestrial region. An indication of the collision conditions is given by the arrows in Figure 23.3. The detailed way that the planets approach each other is also important, in particular the *offset*, illustrated in Figure 23.4. In the calculation to be described the offset is 4×10^4 kilometres.

23.4 The Composition of the Colliding Planets

Another important consideration that will influence the outcome of a planetary collision is the composition of the planets. They were newly formed from cold material consisting of a mixture of hydrogen and helium within which there were tiny micron-sized grains. These grains were of three basic types — iron, silicates and volatiles, the last consisting of materials such as water (H_2O), methane (CH_4), ammonia (NH_3), carbon dioxide (CO_2) and other more complicated molecules, all in an icy solid form because of the low temperature. The silicate and iron grains are expected to have been coated with ices. The amounts of the substances contained in the grains present in the proposed colliding planets are given in Table 23.1. Much of the icy material contained hydrogen and what is of interest here is its deuterium to hydrogen ratio, D/H.

Solar-system ratios of D/H vary from a low of 2×10^{-5} in Jupiter to a high of 1.6×10^{-2} in Venus. The Jupiter value is also that for the Universe at large since Jupiter has retained all the material that originally went into its formation. Comets, meteorites and the Earth all show values of D/H intermediate between those of Jupiter and Venus.

By analysing the infrared radiation from various cold sources within the galaxy it is possible to determine not only the types of chemical molecules that are present but also the ratios of D/H in them. Observations since 2001 have shown remarkably high D/H ratios in some molecular species in cold dense clouds and in newly-formed protostars within them. Ammonia in the cold dense cloud L134N has a ratio $NHD_2/NH_3 = 0.005$ and in a protostar 16293E that ratio is 0.03. For formaldehyde the ratio D_2CO/H_2CO is in the range 0.01 to 0.4 in a number of protostars. Deuterated methanol, CH_3OH, in the protostar IRAS 16293–2422 actually exceeds in quantity that containing just hydrogen. The observed ratio of HDO/H_2O tends to be smaller, about 0.004, but in the inner core regions of IRAS 16293–2422 a ratio 0.03 has been observed.

Overall a cold dense cloud has the normal cosmic ratio $D/H = 2 \times 10^{-5}$ but deuterium has been concentrated in icy grains by a process called *grain surface chemistry*. When a deuterium atom falls on an icy grain it swaps places with a hydrogen atom in a molecule because the substitution gives a slightly more stable molecule. It is a small effect but over a long period of time it leads to a large enhancement of the deuterium in the grain. When a protostar is formed from cold dense cloud material, the deuterium-rich grains migrate towards the centre. As the protostar evolves towards being a star the high D/H ratio at its centre persists for some time although, eventually, when the molecules dissociate into their elements due to the high stellar temperature, the deuterium so produced gradually diffuses to all parts of the star and the deuterium excess disappears. The capture-theory model produces planets either directly from compressed cold cloud material or from a young protostar. In either case parts of the central regions of the protoplanets will be deuterium-rich; a modest estimate of the enrichment has been taken as $D/H = 0.01$.

A planet in the final stages of its evolution will have become differentiated to a great extent. The central core will be predominantly silicate and iron with the proportion of silicate to iron increasing with increasing distance from the centre. Surrounding the central core there will be a region rich in volatiles, which would be in gaseous form because of the high local temperature. In the early stages of evolution of the planet, when it was still cold, grains, including ice grains, would have migrated towards the centre and when the volatiles first melted, and then vaporised, they would have formed a deuterium-rich shell around the dense core.

23.5 Temperature Generated by the Collision Process

When two uniform streams of gas collide head-on, the gas is compressed and heated within two shock fronts (Figure 23.5(a)). These shock fronts move outwards as the collision progresses and within the shock fronts both the density and temperature are enhanced. For gas that is a common astronomical mix of hydrogen and helium, and with each stream moving at 40 kilometres per second, the temperature in the shocked region is of order 150,000 K. The temperature increase comes about because the gas within the shock fronts is stationary and the energy of motion of the gas streams is converted into heat energy.

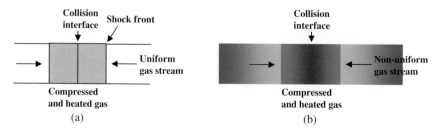

Figure 23.5   Collision of gas streams (a) with uniform gas streams (b) with non-uniform gas streams.

If the density of the gas streams is not uniform but increases with distance from the collision interface (Figure 23.5(b)) then the outcome is quite different. Gas initially compressed near the collision interface, acquiring an initial temperature 150,000 K, is continuously further compressed by the oncoming ever-higher-density material and so heats up further. The greater the increase in density from the gas that first impacts to that arriving later, the greater the temperature of the gas becomes at the collision interface. A major planet has just these conditions, with a density similar to that of air, of the order of 1 kilogram per cubic metre, near the surface and something in the region of tens of thousands of kilograms per cubic metre near the core.

Because of the density distribution, augmented by increasing temperature with distance from the collision interface, a planetary collision would have generated a very high and steadily increasing temperature, which would have penetrated further and further towards the centre of the planets as the collision progressed. The question then arises of whether conditions occurred in which nuclear reactions could have taken place. The factors that govern the rate of any nuclear reaction are the temperature and the densities of the reacting nuclei. The temperature at the centre of the Sun is about 15 million K and this enables nuclear reactions to take place that convert hydrogen into helium although at a modest rate, which is fortunate as the Sun would otherwise have exhausted its hydrogen and moved off the main sequence long ago.

A planetary collision takes place over a short timescale — a few hours — so what we are interested in are conditions that would give *explosive* nuclear reactions. To give such conditions most nuclear reactions require a temperature of hundreds of millions K but with an interesting exception, a reaction involving two deuterium atoms. Two deuterium atoms can react in one of two ways:

$$ {}_1^2D + {}_1^2D \rightarrow {}_1^3T + {}_1^1P $$

or

$$ {}_1^2D + {}_1^2D \rightarrow {}_2^3He + {}_0^1n. $$

In the top reaction the products are tritium and a proton (equivalent to a hydrogen nucleus, with unit charge and unit mass) and the bottom reaction gives helium-3 plus a neutron, with zero charge and unit mass. Tritium and helium-3 then engage in further nuclear reactions. If the deuterium, or material containing deuterium, is at a high density then this reaction begins to be explosive at just above 2 million K. Such conditions can occur in a planetary collision and when it does so there is a nuclear explosion.

23.6 Modelling the Collision

The planets in Table 23.1 were modelled using SPH (Section 19.2) with the assumption that segregation by density had not gone to completion. They were represented by four layers — a core with slightly more silicate than iron, a mantle, mostly silicate but with some residual iron, a region of deuterium-enriched hydrogen compounds, referred to as 'ice' regardless of its physical state, and finally, a hydrogen-plus-helium atmosphere. The SPH particles were more concentrated in the central regions so that the core, mantle and ice were represented by many SPH particles. This was desirable since the outcome for the central material was the essence of the calculation and with many particles its behaviour could be followed in detail. The result of the SPH calculation is shown in Figure 23.6. The frames of the figure show:

(a)   The starting point for the simulation $(t = 0)$ just before contact. Distortion of the planets is not discernable.
(b) $t = 501$ s. Enyo is highly distorted and material is being sprayed out sideways from the collision interface.
(c) $t = 1{,}001$ s. This is similar to (b) but the high-density and hot shock interface is close to the ice region of Enyo.
(d) $t = 1{,}501$ s. The shock region has reached the ice of Enyo, the temperature climbs above 2 million K, and within a few seconds D-D nuclear reactions occur.
(e) $t = 2{,}003$ s. The atmosphere and some of the core and mantle material are being propelled outwards.

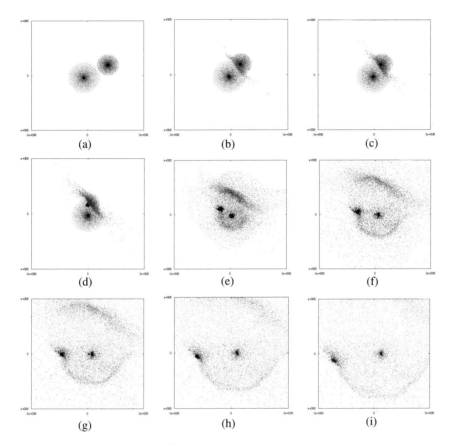

Figure 23.6 The progress of the planetary collision. (a) $t = 0$, just before contact, (b) $t = 501$ s., (c) $t = 1,001$ s., (d) $t = 1,501$ s., (e) $t = 2,003$ s., (f) $t = 2,511$ s., (g) $t = 3,004$ s., (h) $t = 3,502$ s., (i) $t = 4,005$ s.

(f) $t = 2,511$ s. There is further expansion of the material but core + mantle residues are seen for each of the planets.

(g) $t = 3,004$ s. More expansion with residues moving apart.

(h) $t = 3,502$ s. More expansion with residues moving apart.

(i) $t = 4,005$ s. More expansion with residues moving apart.

The D–D reactions raise the temperature to a level at which other reactions, involving heavier elements, can take place. By the end of the simulation the temperature in the ice region has risen to several hundred million K. What we have at the end of the calculation are two iron-plus-silicate residual cores, with masses of the same order as those of the Earth and Venus. A straightforward dynamical calculation shows that the colliding planets, initially on fairly extended partially evolved orbits around the Sun, could give two solid residues with orbits that rounded off in the terrestrial region of the Solar System. It is proposed that Venus and Earth are formed from the residues of Enyo and Bellona, respectively.

A feature of Earth and Venus, explained by the collision hypothesis, is that they both have a ratio of iron to silicate considerably higher than that estimated for the Universe at large. The surviving cores shown in Figure 23.6, which form the larger terrestrial planets, are residues of the much larger silicate-plus-iron cores of the original major planets. The outer parts of the original cores that were stripped away were silicate rich and what was left was iron rich — as are the Earth and Venus.

The planetary-collision scenario explains the existence of the larger terrestrial planets, although the primary capture-theory mechanism only gives major planets. If that were all a planetary collision could explain then it might seem to be a rather *ad hoc* proposal for explaining two terrestrial planets and nothing else. It would clearly be much more convincing if it could also explain other solar-system features — for example, the other two terrestrial planets. In fact, the planetary collision hypothesis offers explanations for many, apparently disparate, features of the Solar System — as we shall see.

Chapter 24

The Moon — Its Structure and History

24.1 Ideas About the Origin of the Moon

The Moon and its main characteristics were described in Section 16.6 and it is clear that it poses many questions about how its surface features were formed. It has hemispherical asymmetry with the near side dominated by large mare basins while the far side consists almost completely of highland regions. The hemispherical asymmetry is linked to the crust thickness, which is thinner on the near side although, from theoretical considerations, it should be thicker.

The very presence of the Moon, so anomalously large compared to its parent planet, is a mystery that has attracted a great deal of attention. The three main contending theories that have been put forward to explain its presence are:

(i) It was formed in association with the Earth.
(ii) It was formed separately from the Earth but was captured by it.
(iii) The Earth was struck by a large body and debris from the collision, orbiting the Earth, came together to form the Moon.

Two mechanisms have been suggested for the first of these possibilities. In 1873 Édouard Roche (1820–1883), a French scientist and expert on celestial mechanics, suggested that the Earth and the Moon were both accumulations of smaller bodies that happened to form in a binary association. This idea is consistent with the Solar Nebula Theory, where both bodies could form from planetesimals. Another idea fitting with explanation (i) was advanced by George Darwin (1845–1912), an eminent mathematician and astronomer of his day, and the son of Charles Darwin of *On the Origin of Species* fame. Darwin envisaged a quickly-spinning molten Earth that became unstable under the action of solar tides so that a Moon-size chunk separated from the Earth. This idea was stimulated by the fact that the Pacific Ocean represents an almost complete hemisphere of the Earth that is denuded of continental crust material and also that the Moon is receding from the Earth, by the process described in Section 20.5. Darwin estimated that at some stage the Moon would have been very close to the Earth and hence may have been derived from it. The problem with this explanation is that, if all the mass and angular momentum associated with the present Moon was absorbed into the Earth, it would still not be spinning fast enough to become unstable, even with the help of the Sun's tidal influence.

Idea (ii) considers that the Earth and the Moon were both in independent heliocentric orbits and approached each other in such a way that the Moon was captured by the Earth. In Section 21.2 the problems associated with capture were explained; it would have been necessary somehow to remove energy of motion from the Earth-Moon combination. A capture process cannot be ruled out but it is so unlikely that it would be reasonable to consider it only as a last resort.

The final possibility has been worked on extensively by the American astronomer W. Benz and his colleagues and is widely supported. An SPH simulation of the process is shown in Figure 24.1. A Mars-mass body strikes the Earth obliquely, and debris coming from both bodies forms the Moon, seen most clearly in frame (h) of the figure. There are no serious problems with this model although there are a few minor ones, since models tend to give 'Moons' that are too massive and with less iron than is inferred from the Moon's known density. Another concern is that it postulates a different origin for the Moon than for all the other larger solar-system satellites, yet the Moon fits comfortably between Io and Europa in both mass and

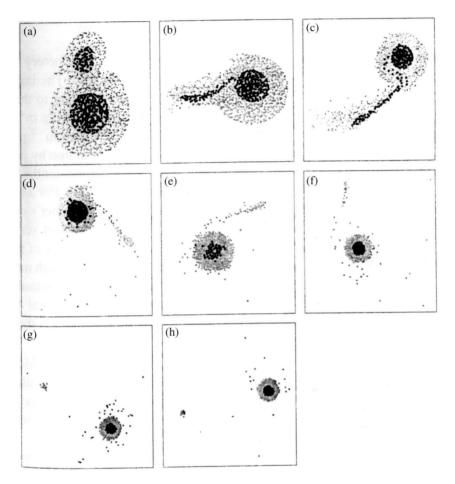

Figure 24.1 Stages in the formation of the Moon by the single-impact process. Grey regions represent silicate and black regions are iron.

density. However, some discomfort with the rather *ad hoc* nature of the postulated mechanism does not detract from its plausibility. No model is ever perfect and it is rarely possible to explore a full range of parameters for a model in order to get best agreement with observation. Whether it actually happened or not, it would be possible to produce an Earth satellite by an impact and such a satellite might resemble the Moon.

24.2 The Planetary Collision and the Moon

In the postulated planetary collision the planetary masses were, respectively, nearly twice that of Jupiter and more than that of Saturn. From the reasoning given in Chapter 21 the colliding planets would certainly have had satellites. The number of possible outcomes for satellites following the collision is limited and now listed:

(a)   The satellite could have been destroyed by a collision with a large piece of debris. Unless the satellite was very close this is unlikely, but possible.

(b)   The satellite could have gone into an independent heliocentric orbit. After a few hours the bulk of the planets had dispersed and the gravitational pull of the residue of the planet could have become insufficient to retain the satellite.

(c)   The satellite could have escaped from the Solar System. This outcome is related to (b) and depends on the speed of the satellite relative to the Sun.

(d)   The satellite could have been retained by the residue of its original parent planet or have become a satellite of the other residue.

We are now going to consider outcome (d) to describe the origin of the Moon. Comparing its mass relative to the larger satellites of Jupiter and Saturn, the Moon could have been a satellite of either of the colliding planets. Computation shows that the probability of retention by the original planet residue is greater than that of capture by the other residue so here it will be taken as a satellite of Bellona, whose residue now forms the Earth. At any reasonable distance from the planet its orbital period would have been of the order of a day at least — for example, Io, the innermost Galilean satellite has an orbital period of about 42 hours. The Moon would have kept one face directed towards its planet and during the hour or two when the process of planetary collision was taking place, when most debris was being created, the orientation of the Moon relative to the collision region would have changed little. This means that just one hemisphere of the Moon's surface was exposed to debris and we must now consider the effect of that debris on the Moon's surface.

Experiments on high-velocity impacts of bodies, carried out at NASA's Ames Research Center in California, show that the debris ejected sideways from a collision interface, as seen in Figure 23.6, has speeds up to three times the impact speed of the bodies. This would imply speeds of up to 240 kilometres per second relative to the colliding bodies, although most debris would be moving at considerably less speed. What would have happened when debris from the collision struck the surface of the Moon? The escape speed from the Moon is 2.4 kilometres per second, meaning that any object leaving the surface of the Moon at that speed or greater would be lost. Again, any object coming from a large distance and falling on the Moon must do so with *at least* the escape speed. Figure 24.2 shows the possible outcomes when a projectile strikes the Moon. In Figure 24.2(a) the object falls on the surface with just over the escape speed. It breaks up and shares its energy with greater than its own mass of surface material. All the material that flies up from the surface has less than

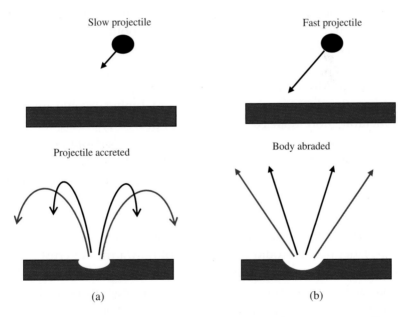

Figure 24.2 The effect of slow and fast projectiles falling on to a satellite.

the escape speed and so falls back onto the surface. The result is that the projectile is *accreted*, much as meteorites, slowed down by their passage through the atmosphere to which they lose energy, are accreted by the Earth. In Figure 24.2(b) we see what happens when an object falls onto the surface with a speed considerably greater than the escape speed. Much of the material that is broken off the surface by the projectile now has more than the escape speed and so escapes from the Moon; in this case the result is that the surface is *abraded*, i.e. loses material.

If the average speed of projectiles falling on the Moon's surface was, say, 100 kilometres per second then, if that energy was shared equally with surface material, theoretically it could abrade up to 1,600 times its own mass off the surface. Actually much of the projectile's energy goes into breaking up and heating surface material so the amount of abraded material is probably closer to 100 times the projectile mass. To thin the crust of one hemisphere of the Moon by 25 kilometres requires the removal of about 10^{21} kilograms of crust (1.4% of the mass of the Moon) and, conservatively, this would require 10^{19} kilograms of projectile material to fall on the Moon. While this may seem a large mass it is only about twice the mass of the Earth's atmosphere. It is also about one ten-millionth of the mass of the solid debris coming from the collision and a body with the radius of the Moon at, say, 1,000,000 kilometres from the collision (less than the distance of Ganymede from Jupiter) would intercept considerably more than 10^{19} kilograms of debris. There are too many unknowns to do precise calculations — for example, from Figure 23.6 it appears that the debris is concentrated in some directions — but rough back-of-envelope calculations of the type given above confirm the plausibility of this mechanism for stripping away part of the Moon's crust.

The tidal forces on the Moon when associated with its original major-planet parent would have given not only a thicker crust on the near side, as shown in Figure 16.17, but would also have led to change in its overall shape and internal rearrangements of the core and mantle. The present slightly pear-shaped Moon, together with the changes in internal structure, ensured that when the Moon settled down after the collision, it was the abraded face that faced its depleted parent.

From the collision event that explains the thinned near-side crust, all the other major surface features of the Moon follow. However, to understand this we first need to consider the thermal implications of the way that satellites form. In Section 22.1 it was explained that as a body grew by accreting material so the temperature of the added material would steadily rise, corresponding to the increasing speed at which it strikes the surface. At a radius of about 1,100 kilometres added material will be at its melting point so that when a Moon-size body (radius 1,740 kilometres) is newly formed it is solid on the inside and molten on the outside. The radiating surface would cool quickly so a solid crust would form, but this new crust would constantly be disrupted by convection currents within the fluid region, resulting in strong volcanism that would bring new material to the surface that, in its turn, would cool and solidify. Eventually, a stable crust would form; a representation of the thermal profile at this stage is shown in red in Figure 24.3. As time progressed the solid crust would have thickened and the molten region would have migrated in towards the centre (green line in Figure 24.3).

With a sufficiently thick crust, any large projectile falling onto the surface over the next few hundred million years would excavate a

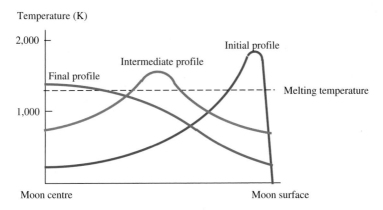

Figure 24.3 A sequence of thermal profiles for the Moon. The molten regions are those above the melting temperature.

large basin and also form cracks beneath the basin. If the cracks extended down to the molten region, then magma would have risen though the cracks to fill the basin, so creating the prominent lunar maria (plural of mare) on the near side of the Moon. On the far side with its thicker crust, for the most part the cracks did not penetrate to the molten region and so there were many excavated basins, but only one mare of appreciable size — Mare Moscoviense.

The dates of the maria outflows are between 3,200 and 3,900 million years before the present. These do not represent the dates of formation of the basins. Volcanism occurred in episodic fashion and earlier flows covered later ones. In some maria one can see 'flow-fronts' — ridges corresponding to the limits of a magma flow. With time, the magma that came to the surface was cooler and more viscous so that later flows went less far than earlier flows thus creating a series of flow-fronts, each closer to the source of the eruption than the flow that preceded it. Smaller projectiles produced craters, particularly abundantly in the highland regions. There are fewer in mare regions since earlier craters were covered by volcanic outflows.

It will be seen that the most important features of the Moon's surface can be explained in terms of the outcome of the planetary collision. It can also be seen that the Moon is no different from any other satellite in terms of its origin. It was produced in the same way as other satellites in association with a major planet. The question should not be 'Why is the Moon so large compared to the Earth?' but rather 'Why is the Earth so small compared to the Moon?' and the answer given here is that it is the residue of a much larger body. The mystery lies not in the nature of the Moon but in the nature of the Earth.

Chapter 25

The Very Small Planets — Mars and Mercury

A planetary collision gives a plausible explanation for the origin of the Earth and Venus as major-planet residues, but we must look for another explanation for the origin of the remaining terrestrial planets, Mars and Mercury. To compare these latter planets with other sizable solid solar-system bodies, Figure 25.1 shows the densities and masses of the terrestrial planets and some rocky and icy satellites shaded according to the type of body.

The horizontal dashed line is at the centre of the largest gap in the density range and on this criterion Mars seems more akin to larger satellites than to the other three terrestrial planets. The vertical line is at the centre of the largest gap in the run of masses (plotted on a logarithmic scale so that each of the major divisions along x represents a factor of ten). On this basis Mars and Mercury seem more akin to the satellites than to the larger terrestrial planets. On both criteria Mars seems more comfortable as a large satellite than as a small planet and we consider the possibility of a satellite origin.

25.1 The Origin of Mars

Mars has four times the mass of Ganymede, the most massive satellite, which might be considered as counter-indicative of a satellite origin. However, one of the colliding planets had twice the mass of Jupiter so that, with much more mass available in the disk surrounding the

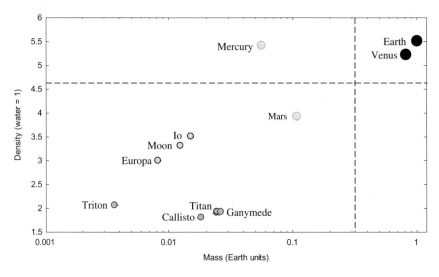

Figure 25.1   The densities and masses of larger solid bodies in the Solar System.
● Larger terrestrial planets, ◍ Smaller terrestrial planets, ○ Rocky
satellites, ⊙ Icy satellites.

planet, a more massive satellite might be expected. This argument is
suggestive, but not conclusive, so we seek other indications that Mars
was a one-time satellite of the larger colliding planet that went into an
independent heliocentric orbit and eventually settled down where it
is now.

If Mars had been a satellite then its hemispherical asymmetry
(Section 15.3 and Figure 15.9) could be explained in the same way as
that of the Moon. Since Mars considerably exceeds the size of body
that would be molten when it accumulated, the solidified crust that
initially formed on Mars would have been floating on a low-viscosity
fluid sea of molten magma. Now, when the crust of one hemisphere
was abraded, and possibly completely removed, what was exposed was
the denser mantle material below while in the unaffected hemisphere
there remained a floating island of less-dense crustal material. This
contrast between the two hemispheres is partially disturbed by the
Hellas Basin, caused by a huge projectile that smashed its way though
the crust of the southern highlands.

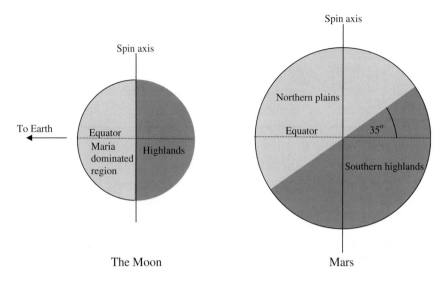

Figure 25.2 Hemispherical asymmetry on the Moon and Mars.

There is an important difference between the hemispherical asymmetries of the Moon and Mars, apart from the difference in the extent of abrasion. For the Moon the line dividing the hemispheres is perpendicular to the equator whereas for Mars this line is at 35° to the equator. These two configurations are illustrated in Figure 25.2.

The Moon as a regular satellite of the larger colliding planet would have orbited in the plane of the planet's equator with its spin axis perpendicular to its orbital plane. Tidal forces would have moulded it into a pear-shape with the pointed end towards the collision; the tidal force due to the planet residue, the Earth, would then have acted on the Moon in such a way as to preserve that configuration, which is why the damaged hemisphere of the Moon now faces the Earth. By contrast, Mars went into an independent heliocentric orbit and was not subjected to tidal effects. In its initial state, just as for the Moon, the spin axis would have passed through the plane of asymmetry. Now we describe the process of *polar wander*, by which surface features move relative to the spin axis — related to *continental drift* by which continents on Earth have moved relative to the spin axis (Chapter 34). Early Mars would have had an extensive molten interior with the fluid boundary just below the solid surface material, which would have been the surviving solidified crust and solidified magma where the crust had been removed. The material in a rotating partly-fluid body generates velocity gradients within the fluid, and viscosity effects turn some of the energy of motion into heat energy that gets radiated away. For this reason the body loses kinetic energy but, as it is isolated, it cannot lose its angular momentum. It can be shown theoretically that, to satisfy the requirement that the kinetic energy reduces while the angular momentum remains constant, the material of the body must rearrange itself so as to move away from the spin axis. The surface of Mars has several features that give considerable deviation from a spherical form. Taking the average height of the highland region as a reference level, both the Northern Plains and the Hellas Basin are large negative features below that level. There are a number of positive features that are higher than the reference level, for example, the volcano Olympus Mons shown in Figure 15.8 and some upland regions — the Tharsis Uplift, Elysium Plains and the Argyre Plain — that are several kilometres above the mean level. Over time Mars rearranged its surface material by the equivalent of continental drift, whereby the whole solid surface layer slid over the molten mantle material beneath. From the theory given, the movement should have positioned positive regions as far from the spin axis as possible while getting negative regions as close to the spin axis as possible. Calculations show that the present configuration is very close to this ideal.

It is probable that Mars, like the Galilean satellites from Europa outwards, would have had an icy covering above the solid crust before the planetary collision occurred. This would have melted and then vaporised as a result of the subsequent volcanism forming a water-rich atmosphere; water vapour acts as a greenhouse gas and this would have raised the temperature of Mars. As the planet cooled this would have given a period of a water-dominated climate with clouds and rain, so producing the fluvial features — dried up river beds — that are observed in images of the surface (Figure 15.7). Eventually much of the atmosphere would have been lost since, when the temperature was higher than it is now, Mars' gravity would have been insufficient to retain it. However, enough water was retained so that, when Mars cooled to its present state, ice was formed that is now the permanent constituent of the polar caps and also exists below the surface in other regions of Mars.

25.2 The Origin of Mercury

The mass and density of Mars, as given in Figure 25.1, plus the interpretation of its surface features and compositions, gives a self-consistent picture of a satellite origin. For Mercury the indications are contradictory — it has a satellite-like mass but a density similar to those of the larger terrestrial planets. However, the collision model gives no possibility of another dense body forming directly unless one of the residues split into two, which seems unlikely from the results of the modelling displayed in Figure 23.5. A comparison of the internal structures of Mars and Mercury is given in Figure 25.3. The iron cores are similar in size and Mercury resembles a Mars-like body that has had much of its crust and mantle removed. This relationship has been noted in the past and it has previously been suggested that Mercury was once similar to Mars but had outer parts of it stripped away by some catastrophic event. What we do here is identify that catastrophic event.

Since the surfaces of the Moon and Mars have been interpreted in terms of abrasion by debris from a planetary collision, it is an obvious extension to consider the outcome for a very close satellite that was so heavily exposed to debris that the majority of its crust and mantle

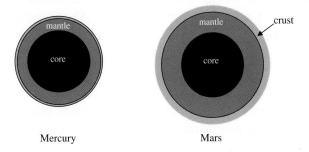

Figure 25.3   The internal structures of Mercury and Mars.

on one side was removed. A body so deformed would rearrange itself under its self-gravitational forces into a more spherically-symmetric form. The rearrangement of the mantle to envelop the core would have released a considerable amount of gravitational energy so that the whole body would have been in a high-temperature, low-viscosity molten state. Mercury formed in this way would *not* have hemispherical asymmetry. As Mercury cooled so a solid surface layer formed that eventually was able to record the impact history in the form of craters. The cratered surface of Mercury has been likened to the highland regions of the Moon but there are important differences. The density of craters is less on Mercury, indicating that for some time following the collision, when the density of debris was greatest, the surface crust was not thick enough to record craters — it would be like throwing a stone though thin ice that refreezes to hide the original damage. Another difference is that on Mercury the spaces between the craters are occupied by volcanic material, which is not a feature of the lunar highlands.

25.3 The Moon, Mars and Mercury — a Summary

To summarise — the overall conclusion is that there were at least three large satellites of the larger colliding planet. Mercury and Mars both originally had a mass about four times that of Ganymede. The third satellite was the Moon. The Moon was the most lightly abraded, so that the crust of one hemisphere was thinned to the extent that the large basins excavated by later large projectiles could fill with magma from below to give maria. Tidal locking of the Moon to the Earth kept the abraded side pointing earthwards. The next most heavily abraded satellite, Mars, had most of its crust stripped away in one hemisphere leaving the remaining crust like a continent floating on a sea of magma. Polar wander, dictated by the need to get mass as far from the spin axis as possible, gave the arrangement of hemispherical asymmetry relative to the spin axis now observed. The final satellite, Mercury, was so heavily abraded that a large proportion of its crust and mantle was lost from one hemisphere. Subsequent reconfiguration gave a spherically symmetric body of high density but low mass.

Chapter 26

Smaller Bodies of the Solar System

26.1 Asteroid Formation

Meteorites are mostly fragments from colliding asteroids, although a few almost certainly come from Mars and the Moon. The distinct types of meteorites — stones, irons and stony-irons — suggest that the original sources — asteroids — were differentiated bodies. The planetary collision offers an obvious source of asteroids, the different types coming from different regions of the differentiated planets. Most material from the central regions of the colliding planets was thrown out as debris; only small residues of these regions ended up as the Earth and Venus. Some debris would initially have been in liquid form but would quickly have solidified and compacted to form asteroids. Stones and irons would have come from the mantles and cores of the central regions. The stony-iron pallasites represent an orderly assembly of stone and iron from partially-differentiated regions where stone and iron coexisted, since differentiation in the colliding planets was incomplete, while the structure of mesosiderites suggests that stony and iron material thrown out from different parts of the planets spattered together in a violent fashion.

The postulated original mass of debris is much greater than the estimated combined masses of all the existing small-scale solar-system material, which is of the order of one Earth mass. The collision took place within the terrestrial region of the Solar System and material would have been thrown into a wide range of paths, some elliptical with high eccentricities and semi-major axes from a few to several thousand astronomical units, and some that left the Solar System altogether. The motions of the debris that went into elliptical orbits were disturbed by a myriad of bodies. Debris with orbits that repeatedly brought them into the inner Solar System would have had little chance of surviving to the present time. At some stage they would have interacted with a major planet, either by collision, in which case they would be absorbed and disappear, or by being deflected from their original paths, which could either throw them into a new elliptical orbit or could project them out of the Solar System. If they went into a new elliptical orbit then they would remain at risk because their orbits would again repeatedly return them into the region of the major planets.

The structure of the central region of a colliding major planet, differentiated by density, gave silicate materials progressively more impregnated with volatile material with increasing distance from the centre. Volatile material on a terrestrial planet — the Earth for example — appears as the oceans and atmosphere but in a major planet the same material can be considered as either the lower reaches of the atmosphere or as an outer layer of the central core. The mechanics of the explosion ensured that the debris from inner regions of the core was more constrained in its outward motion and hence travelled out less far. Thus iron debris, and silicate debris with little volatile content, would tend to be concentrated within the inner Solar System; this is potential asteroid material. The silicate material furthest out from the centre of the planet, with the highest inventory of volatiles — potential comet material — would have been thrown much further out. Now we consider how such debris fared after its formation.

In order to survive and to be safe from catastrophic events an asteroid would somehow have had to attain an orbit that kept it away from the planets — especially the major planets with their strong gravitational fields. Immediately following the collision, the asteroids would have interacted gravitationally with all other bodies, including planets, and by collision with each other. Once the planets had settled down into their final orbits, debris remaining well within the orbit of Jupiter would have a strong chance of survival, especially if its orbit kept it between Mars and Jupiter, the region now referred to as *the asteroid belt.* Some asteroids outside this safe zone also managed to survive, for example, Apollo, an Earth-crossing asteroid, and Chiron, the orbit of which keeps it mostly between Saturn and Uranus.

26.2 Comets and the Kuiper Belt

An interesting modern solar-system discovery is the Kuiper Belt (KB), briefly described in Section 17.5. Most constituent bodies of the KB are believed to be similar to comets, the debris thrown further out after the planetary collision; inner KB objects, perturbed by Neptune, are believed to be the source of short-period comets. We must now consider how it is that KB objects ended up in orbits that were completely outside the present planetary region.

The planetary collision took place early in the life of the Solar System when planetary orbits were still extended and the planets at that time would have ranged out to several hundreds of astronomical units. Because the colliding planets were moving rapidly close to the general plane of the planetary orbits, which were close to being coplanar although not precisely so, the motions of the retained debris — that which did not leave the Solar System — would have also tended to be close to the general plane, where 'close' in this context means within 30° or so. Debris close to the general plane was then orbiting on paths that took them to the same outlying regions as were occupied by the extended orbits of major planets. Because of the mechanics of planetary motion, these early planets in highly eccentric orbits moved slowly near aphelion and spent most of their time at large distances from the Sun. Some proportion of the debris would have gravitationally interacted with these distant major planets and their paths would have been changed to a greater or lesser extent. Figure 26.1 shows a possible outcome of an interaction between a potential comet or KB object and a major planet in a region beyond the present orbit of Neptune. In the figure the planet is seen on an extended orbit which is still evolving. The initial comet orbit is extended with a small perihelion; the comet is seen passing close to the planet, the gravitational effect of which swings it into a new orbit. This orbit has a perihelion outside the orbit of Neptune,

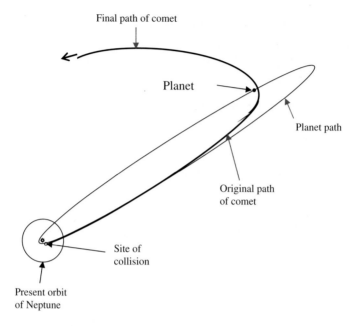

Figure 26.1   The interaction of an asteroid on an extended, highly eccentric orbit with a planet.

the outermost major planet. If this comet then survived for the limited period that it took for the planetary orbits to completely evolve then, thereafter, it was safe and could survive indefinitely.

Such a scenario explains the existence of KB objects and also suggests that the KB may extend outwards to a great distance, although all we detect is its innermost region.

26.3 The Oort Cloud

The idea of an inner belt of comets has been proposed by a British astronomer, Mark Bailey, as a potential source of the comets that occupy the Oort Cloud, a cloud of comets existing mostly at distances of tens of thousands of astronomical units, with the perihelia of some of them much further in although well outside the planetary region.

When the Oort Cloud was discovered, one problem it raised was how it survived. Knowing the separations and speeds of stars in the solar environment it is estimated that within the lifetime of the Solar System about 4,000 stars will have passed within 30,000 astronomical units from the Sun, about 1,000 within 15,000 astronomical units and 4 within 1,000 astronomical units. The problem was how the Oort Cloud could have survived the predation of these stellar invaders, and the problem is exacerbated by the existence of other potential disturbing bodies, such as Giant Molecular Clouds (GMCs), dense clouds of material about 100 light years across with total mass about one million solar masses. They are lumpy in structure and, if the Solar System passed through a GMC, bodies at the distance of Oort-Cloud comets would, for the most part, become detached from the Sun. It is probable that the Solar System has passed through up to four GMCs during its lifetime.

With an inner reservoir of comets, as Mark Bailey suggests, the difficulty of explaining the survival of the Oort Cloud is resolved. A star passing at a distance of 10,000 astronomical units would certainly remove many Oort-Cloud comets from the Solar System but it would also perturb some inner-reservoir comets outwards to replace them. In this way, until the inner reservoir becomes exhausted, the Oort Cloud can survive.

26.4 The Dwarf Planets

Before 2006 Pluto was regarded as the outermost planet of the Solar System. However, in 1992 astronomers began to detect substantial bodies in the KB. In 2006 a body, Eris, was found with a satellite (Figure 26.2), thus enabling its mass to be found, which was 27% greater than that of Pluto. In 2006 a meeting of astronomers was held to decide what to do — was Eris to be a tenth planet or was Pluto to be demoted in some way? The decision was to demote Pluto and to set up a new class of objects, *dwarf planets*, defined as being bodies in heliocentric orbit and large enough to be in spherical form due to self-gravitational forces, but not including the terrestrial and major planets. By 2011 there were five designated dwarf planets, listed in Table 26.1.

The dwarf planets are all within the range of size and mass of some of the smaller regular satellites of the Solar System

Figure 26.2   The dwarf planet Eris and its satellite Disomnia (Hubble Space Telescope).

Table 26.1   Characteristics of the dwarf planets (α is semi-major axis, ε is orbital eccentricity and ι is inclination).

Body	α	ε	ι	Diameter (km)	Mass (10^{21} kg)	Satellites
Ceres	2.77	0.080	10.6	975	0.95	0
Pluto	30.48	0.249	17.1	2,306	13.05	3
Haumea	43.34	0.180	28.2	1,150	4.2	2
Makemake	45.79	0.159	29.0	1,500	4?	0
Eris	67.67	0.442	44.2	2,400	16.7	1

and can comfortably be interpreted as satellites that escaped from the collision. Ceres could have survived in the same way as did the asteroids between Mars and Jupiter. Similarly Eris, Haumea and Makemake could have interacted with a planet as shown in Figure 26.1 to reach their present positions. Pluto is somewhat different in origin, as we now explain.

26.5 The Relationship of Pluto and Triton to Neptune

The orbits of Pluto and Neptune are shown in projection in Figure. 26.3. The orbit of Pluto at perihelion is just inside that of Neptune and this relationship has often been taken to indicate some connection between the two bodies in the past. A possible explanation of the relationship is that Pluto, a satellite of a colliding planet, approached Neptune coming from the inner Solar System and was swung by the gravitational effect of Neptune into its present orbit. Computation shows that such a scenario, similar to that shown in Figure 26.1, is quite possible. However, an alternative scenario not only explains the orbit of Pluto but, in addition, provides explanations for other peculiar solar-system features. The first of these is the retrograde orbit of Neptune's satellite Triton, which cannot possibly be a regular satellite. The second is that the small dwarf planet Pluto has three known satellites. Pluto itself has a diameter of 2,306 kilometres and a mass about one-fifth that of the Moon. The diameter of its largest satellite, Charon, is 1,212 kilometres, more than a half that of Pluto, and its mass is about 12% that of Pluto. If the Moon is a large satellite compared to the Earth then Charon is extremely large compared to Pluto. The other two satellites of Pluto, Nix and Hydra, are much smaller, with diameters of order 50 kilometres (Figure 26.4).

We now consider an alternative scenario in which Pluto was not a satellite escaping from the collision but was originally a regular

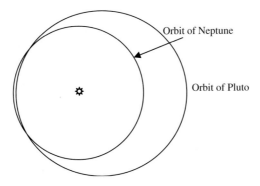

Figure 26.3   The orbits of Pluto and Neptune.

Figure 26.4   Pluto and its satellites (H. Weaver (JHU/APL), A. Stern (SwR) and the HST Pluto companion research team).

satellite of Neptune, and Triton was a satellite of a colliding planet in an extended highly eccentric orbit. Triton, coming inwards from the aphelion of its orbit, struck Pluto a glancing blow, adding energy to it that removed it from Neptune's influence into its present heliocentric orbit and also gave Pluto its retrograde spin. The sideswipe on Pluto broke off fragments that became its satellites — a substantial fragment that became Charon and two small ones that became Nix and Hydra. Triton itself lost energy in the collision and slowed down to the extent that it became trapped within Neptune's gravitational field, so becoming a satellite in a *retrograde orbit*. The closely-circular present orbit of Triton, slowly spiralling in towards Neptune is a consequence of tidal forces acting on a retrograde orbit — the opposite of the Earth's effect on the directly orbiting Moon, which causes it to retreat from the Earth with an increasing orbital eccentricity. The initial state before the interaction between Triton and Pluto is illustrated in Figure 26.5. Lest it be thought that this is just a hand-waving description of a scenario for the Neptune-Triton-Pluto system, without foundation other than a certain self-consistency in providing

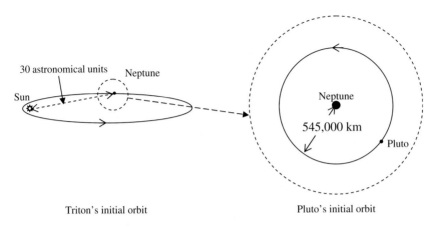

Figure 26.5 The initial orbits of Triton and Pluto before they collided.

the solution to several peculiar solar-system features in a single event, it should be stressed that its plausibility has also been confirmed by detailed mathematical modelling. Another feature associated with Neptune's satellite family is the extreme eccentricity ($e = 0.75$) of the orbit of Nereid. This can be explained by the satellite having been disturbed by Triton, either in the incursion that led to the collision with Pluto or in an earlier incursion in which perturbing Nereid was the only significant outcome.

So far in our description of evolutionary processes in the early Solar System we have referred to eight satellites of the colliding planets — Mars, Mercury, Moon, Eris, Haumea, Makemake, Triton and Ceres. With two large planets involved in the collision there could have been other substantial regular satellites. If so then they could have escaped from the Solar System — or there is the intriguing possibility that there may be one or more of them lurking in the far reaches of the KB, yet to be discovered.

As a final comment — if dwarf planets were redefined as *ex-satellites of colliding planets that are in independent heliocentric orbits and massive enough to take on a spherical form* then Mars and Mercury would be demoted from planetary status to that of being dwarf planets.

Chapter 27

The Origin of Isotopic Anomalies in Meteorites

In Section 22.4, a number of isotopic anomalies in meteorites were described, together with *ad hoc* explanations that had been advanced for some individual anomalies. In the context of a planetary collision the temperature generated is sufficient to trigger deuterium–deuterium nuclear reactions. With the enhanced D/H ratio in the icy regions of the planets due to grain-surface chemistry, the energy from these reactions raises the local temperature to hundreds of millions K, at which temperature reactions involving other nuclei of lower mass take place. In a model explored by Paul Holden and the author, the chain of nuclear reactions, starting with those just involving deuterium, was explored for a compressed mixture of hydrogen-containing volatile materials impregnated with various silicates, such as would be expected in the outer regions of the core of a major planet. Included in the calculation were 568 different nuclear reactions, with the reaction rates, dependent on density and temperature, given by formulae in previously published tables. The result of this calculation explained all the isotopic anomalies described in Section 22.4.[a]

There are a number of factors that have to be taken into account when assessing the outcome of a nuclear reaction chain. For example,

---

[a] A very detailed account of the explanations of isotopic anomalies is found in M.M. Woolfson (2010), *On the Origin of Planets: By Means of Natural Simple Processes* (London: Imperial College Press).

in considering the final amounts of $^{17}_{18}O$ and $^{18}_{8}O$ present it is necessary to add the contributions of the radioactive fluorine isotopes $^{17}_{9}F$ and $^{18}_{9}F$ that decay into $^{17}_{8}O$ and $^{18}_{8}O$ with half-lives of 1.1 minutes and 1.83 hours, respectively. Similar considerations apply in explaining anomalies for other elements. For the individual anomalies the outcomes, allowing for the decay of radioactive products, was as follows.

27.1 Magnesium

Substantial amounts of radioactive $^{26}_{13}Al$ were produced by the reaction chain. Small amounts of this isotope mixed with normal non-radioactive aluminium, $^{27}_{13}Al$, explain the surplus $^{26}_{12}Mg$ found in the white CAI materials in carbonaceous chondrites.

27.2 Oxygen

At the end of the calculation the oxygen that remained was virtually pure $^{16}_{8}O$. There were much reduced levels of $^{17}_{8}O$ and $^{18}_{8}O$ present, the rest having been removed by various reactions. This $^{16}_{8}O$-rich material, when mixed with material that did not undergo nuclear reactions and so retained the normal oxygen isotopic composition, explains all the observed oxygen anomalies. As an example of the kind of results obtained from the reaction-chain computation, Figure 27.1 shows the concentrations of the three oxygen isotopes as a function of increasing temperature as the reaction train proceeds. The amounts of oxygen-17 and oxygen-18 include contributions from fluorine-17 and fluorine-18. At temperatures beyond 600 million K the amounts of oxygen-17 and oxygen-18 have substantially reduced while the oxygen-16 has changed very little.

27.3 Carbon and Silicon

There was copious production of the carbon isotope $^{13}_{6}C$ which, mixed with carbon with the terrestrial ratio of $^{12}_{6}C : ^{13}_{6}C$, gives the range of 'heavy carbon' found in silicon carbide samples. In addition

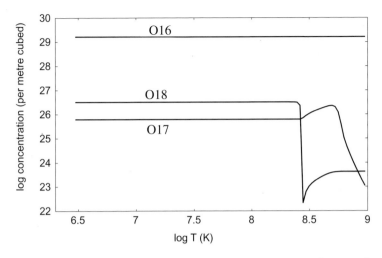

Figure 27.1   The variation of the isotopes of oxygen (including radioactive fluorine contributions) with temperature.

it was found that the relative amounts of the three stable silicon isotopes depended critically on the final temperature in the reaction chain. Silicon derived from different regions of the reacting material, where conditions were different, would have varying silicon ratios that did not correlate with the production of $^{13}_{6}C$. The range of silicon ratios agreed well with those measured in meteorites.

27.4 Nitrogen

The nitrogen in silicon carbide grains is just gas that happened to be around when the grain formed and was trapped in small cavities within the grain. The reaction chain gives the production of the heavier stable nitrogen isotope $^{15}_{7}N$ that, mixing with nitrogen of terrestrial composition, can explain the occasional samples of 'heavy nitrogen'.

In relation to the production of 'heavy carbon' it was mentioned that a great deal of $^{13}_{6}C$ was produced. Another carbon isotope that was produced was the radioactive carbon-14 isotope $^{14}_{6}C$ that would have been included in the silicon carbide that formed. This is the isotope that is used for carbon-dating of historical or archaeological specimens of organic origin. Carbon-14 has a half-life that is just under 5,730 years but it is always present in the carbon dioxide of the atmosphere because it is produced by the action of cosmic rays on nitrogen in the upper reaches of the atmosphere. Thus all living matter contains this isotope in its carbon inventory but, once it dies, the stable carbon remains while the carbon-14 decays and is not replaced. Hence from the ratio of residual carbon-14 to total carbon the time that has elapsed since death can be estimated. For the present purpose the fact that is of interest is that the decay product is the lighter stable isotope of nitrogen, $^{14}_{7}N$. The carbon-14 in the silicon carbide grains completely decays over a period of several tens of thousands of years and the released nitrogen-14 joins the normal nitrogen trapped in grain crevices to give 'light nitrogen'.

27.5 Neon

The nuclear chain produces a considerable amount of radioactive sodium, $^{22}_{11}Na$. This is incorporated into sodium-containing grains that become constituents of meteorites. The decay product, $^{22}_{10}Ne$, is trapped in crevices in the mineral and is released when the mineral is heated, giving the neon-E observations. It must be stressed again that, because of the relatively-localised and short duration of the planetary collision event and its aftermath, the high temperature products of the explosion cool very quickly, in days or even hours, so that the 2.6 year half-life of sodium-22 is not a constraint on this neon-E observation.

It will be seen that the postulate of a planetary collision, and the resultant chain of nuclear reactions, provides straightforward explanations for the isotopic anomalies listed here and negates the need for unrelated, complicated and rather far-fetched explanations for them individually.

Life on Earth

Chapter 28

The Earth Settles
Down — More-or-Less

28.1 The Incandescent Earth

When the colliding planets had disintegrated, the inner cores, which became the terrestrial-planet residues and were at temperatures of several tens of thousands kelvin, ceased to be under high pressure. Under such conditions iron would be highly ionised, i.e. lose many of its electrons, and also become a vapour, at least in the outer regions of the residue. Silicates would have broken down into basic sub-units, as described in Section 22.4, and these would also be in vaporised form. Eventually cooling would reduce the vapour to a liquid state and the Earth would have been an incandescent red-hot liquid ball of iron and silicates with a heaving, boiling surface — a veritable hell. To give an impression of what it must have been like, we must imagine that its whole surface resembled Figure 28.1, which shows part of a lava flow from an eruption of Mount Etna.

The initial orbit of the Earth was eccentric, but it kept mostly within the terrestrial region, and it gradually rounded off to where it is now although, even at present, the orbit does slightly change over long periods of time due to the gravitational influence of the other planets. If there were any intelligent creature in existence that could have surveyed the Solar System at that time, that creature would have seen a Solar System rather different from its present state. The Earth and Venus would have been like small, cool stars, radiating energy

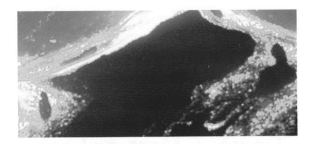

Figure 28.1   An impression of the early surface of the Earth.

from their surfaces and not just reflecting that of the Sun. The Solar System would have been seen as a much more dangerous place than it is now, with large amounts of debris and some ex-satellites of the colliding planets orbiting the Sun and periodically colliding with other bodies and with each other. One of the ex-satellites, the one we now call Triton, ranged out to large distances and led a charmed life every time it approached the region of the planets. A telescope would have shown that Neptune had Pluto as a regular satellite and Mars had no satellites, and, if only the creature were able to make comparisons, it would notice that some of the irregular satellites of the major planets were not yet in place. Yet, despite these major differences, what our intelligent creature would have observed was clearly the Solar System — and it would have been recognised as such by a modern human, although the differences would also be noticed.

Materials were differentiated by density within the Earth's original parent planet and largely remained so in the Earth fragment, but the outer material of the Earth, in particular, would have been stirred up by the violence of the collision. Nevertheless, the differentiation would eventually have re-established itself with an iron core, lighter silicates at the surface and the denser silicates below. The surface regions of the Earth radiated strongly and cooled, and lumps of hot, but solid or plastic, silicates would soon have formed and been tossed around like small icebergs in a stormy sea. As time passed so the total number and size of these solid lumps would have increased while the violence of their motion would have decreased and gradually they would have begun to congeal into large islands of solid material floating on the liquid silicate sea below. This process continued until a recognisable solid crust had formed. At first the crust would occasionally break up under the stresses produced by the movement of the underlying fluid but gradually, as it became thicker, so it became more stable. A similar process, at a much lower temperature, goes on every year in Arctic regions as the sea freezes.

The fluid just below the solid surface was being cooled by conduction of heat through the crust and when cooled its density increased. This meant that it became slightly denser than the material just below it and so it would have moved down to a lower level, being replaced by the less dense material it displaced. This kind of motion of fluid, produced by cooling from above, or by heating from below in the case of heating water in a saucepan, is known as *convection* and is taking place at a very slow rate in the mantle of the Earth at the present time. The rise and fall of different parts of the fluid occurs through the formation of *convection cells*, illustrated in one dimension in Figure 28.2. Since the convection cells bring up heat from below, they accelerate the rate of cooling of the whole body over what it would have been if convection did not occur. Just below the solid surface the fluid is moving horizontally and this will produce a drag force on the surface in the direction of motion, as shown by the black arrows in Figure 28.2. At position A the forces are moving inwards

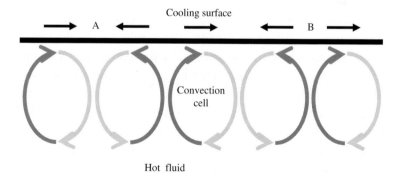

Figure 28.2 Convection cells transporting fluid downwards from a cooling surface and upwards from a hot interior. The black arrows show the direction of drag forces on the surface material.

and are tending to crush surface material while at B the forces are pulling outwards and tending to tear it apart. In general, solids are stronger under compression than under tension so it is the tearing force that will be somewhat more important in terms of disrupting the solid crust. When the early crust formed it would have been too thin to resist rupture and it would have taken a considerable time to thicken to the extent that the forces due to convection could be resisted. The most likely effect of the crushing forces would have been to cause buckling and uplift of the material, like pushing in at the two ends of a sheet of paper. Another effect that would have generated forces on the surface material is the shrinkage of the Earth due to its overall cooling. This would have tended to cause compression forces everywhere on the surface, which would have reacted by wrinkling, something seen on the surface of Mercury due to its shrinkage.

28.2 An Atmosphere Forms

The silicates in the early Earth were impregnated with considerable amounts of volatile matter, the kind that is detected in comets and in some carbonaceous-chondrite meteorites. This would have been retained in the interior of the Earth, where the pressure was high, and would still be present in some quantity when the Earth had formed a thick crust. However, when material rose towards the surface, as shown in Figure 28.2, the pressure on it was reduced and so was its ability to retain trapped volatile material and gasses. We see this effect when opening a bottle of lemonade, which releases the dissolved carbon dioxide gas when the pressure falls to that of the atmosphere. This phenomenon led to the formation of volcanoes through which the released gasses vented to the outside. The gas did not come out alone; just as in present volcanoes it emerged in an explosive fashion and carried with it vast quantities of magma that flowed over the surrounding terrain and also built up the conical peaks that characterise many volcanoes (Figure 28.3). During the earlier stages of the Earth's development, volcanoes were much more abundant and widespread than they are now. As an example we can take Scotland, now bereft of volcanoes and a very stable part of the Earth's surface. Up to about

Figure 28.3 Mount St Helens emitting steam.

three billion years ago the ancient rocks of Scotland were being produced by outflows from volcanoes; indeed, Edinburgh Castle is built on a volcanic plug, a rock formation produced when magma cools and blocks the vent of a volcano. This solidified magma is very durable and when the surrounding softer material was eroded away the plug was left as an isolated hill. What is true for Scotland is also true for many other parts of the world. For example, the *Canadian Shield*, a region covering central, eastern and north-eastern parts of Canada and parts of the United States and Greenland, is a thin layer of soil deposited on igneous rocks laid down by volcanic eruptions. This region is now a stable part of the Earth's crust.

The escaping gases from ancient volcanoes produced both the Earth's early atmosphere and its hydrosphere. Any hydrogen and helium that was still around would quickly be lost; only heavier gasses could be retained. The gasses coming out of the volcanoes would have included:

water (H_2O) carbon dioxide (CO_2) methane (CH_4) ammonia (NH_3)
carbon monoxide (CO) nitrogen (N_2) hydrogen sulphide (H_2S)

with those gasses in the top row being the most common. There would have been traces of other gasses, for example hydrogen chloride (hydrochloric acid, HCl) and some inert gasses like argon and neon, but an obvious characteristic of the early atmosphere is that it was very unlike the atmosphere today. All the listed molecular species would initially have been atmospheric gasses but eventually the Earth cooled to a temperature such that steam could condense to produce water. There would then have ensued a period of heavy precipitation producing large bodies of water in low-lying regions of the Earth. Finally, when the temperature reached levels well below the boiling point of water at the prevailing pressure, the familiar pattern of cloud formation and precipitation in the form of rain would have been established.

The present atmosphere has little resemblance to that which originally existed. It is:

$$\text{nitrogen } (N_2)\ 78\%\ \text{oxygen } (O_2)\ 21\%\ \text{argon } (Ar)\ 1\%$$
$$\text{carbon dioxide } (0.03\%)$$

with traces of other gasses including water vapour. The oxygen, which was not originally present, is the gas that enables advanced life forms, including mankind, to survive. The carbon dioxide that was originally present has mostly disappeared but the small residue plays an important role in maintaining the temperature of the Earth higher than it would be otherwise through the operation of the greenhouse effect (Section 15.3). Without the greenhouse effect the mean temperature of the Earth would be 33° lower than it is now. The present increase in the amount of this gas due to human activity, with a concomitant increase in the greenhouse effect, now constitutes a potential threat to the continuation of most life on Earth.

Apart from that coming from the volatile materials impregnating silicates, the Earth may have acquired more volatiles, which augmented the atmosphere and oceans, from the impact of comets. After the planetary collision there was a considerable rate of bombardment of all the substantial bodies of the Solar System, which can be seen in the numbers of craters that characterise the highlands of the Moon and other inert solid bodies. Water-ice deposits have been found in permanently-shaded regions near the lunar south-pole and these are thought to be due to the impact of comets, although when this happened is uncertain.

The question now arises of how the terrestrial atmosphere changed from its original state to its present state and, in particular, how the oxygen content was produced. It is oxygen that is essential to life but, as we shall see, it is life that enabled the oxygen to be present in the first place.

Chapter 29

What is Life?

29.1 Defining Life

The first of the problems to which man has no solution, and perhaps never will have one, is how the Universe came into being. Once the Universe existed, there followed, by steps that can be readily understood, the complete set of stable elements from hydrogen to uranium, and the ingredients were then available to create all the material entities in the Universe. The most amazing of these entities are those to which the term 'living' can be applied, which encompass a wide range of complexity.

The problem that is intriguing, and to which nobody has yet provided a convincing answer, is that of how life began. But related to that problem is another one — perhaps not so difficult but difficult enough — which is to define what it is that constitutes life. If we look at a dog and a stone we have no difficulty in pronouncing that one is a living entity and the other is a material entity with none of the characteristics of life. But, to take the question one stage further — what are these characteristics of life that the stone does not possess? For example, in Figure 29.1 we show a picture of coral, part of the large structures that form under the sea, such as the Great Barrier Reef to the north of Australia. To a casual observer a coral may seem to be more akin to a rock than to a dog, yet it is a living organism.

To start the process of defining the property that we call life, let us consider the dog and the stone to see what distinguishes them. The stone is easy to deal with. It came into existence as an identifiable

Figure 29.1   Pillar coral (Florida Keys National Marine Sanctuary).

object in its present form some 50 million years ago. It was originally part of a large rock that tumbled down a mountainside and was shattered when a minor earthquake produced a landslide. The fragment from which our stone was produced was then submerged under water for 200 million years. It was rolled this way and that by the moving waters, having its edges ground off by abrasive processes as it slid past other rocks and stones, until eventually it took on a smooth appearance, somewhat like a flattened sphere. The seas retreated and for the last 50 million years our stone has passively rested in its environment of other stones and soil.

Now we consider the dog. It began as an embryo within its mother's womb, just a collection of cells indistinguishable from each other but rapidly multiplying. As the number of these embryonic stem cells increased they began to organise themselves into organs that would perform different functions within the creature they were destined to form — heart, lungs, brain, limbs, and so on. The puppy was born as part of a litter of six, with its mother a Yorkshire terrier and its father of the same breed. It was a tiny bundle of protein and bone, recognisably resembling its parents but completely helpless. However, it was born with some instincts, the most important one being to seek its mother's teats so that it could take in her milk as its first source of food. As the puppy grew so, through experience, it learned the skills that it needed successfully to survive as an adult. After weaning, it consumed solid food, mostly protein but with some carbohydrate as well. It became a subsidiary member of a human family with which it formed an affinity. Its human owner trained it to carry out simple tasks when certain sounds were made — for example, to sit or to roll over. When it was three years old, it was introduced to another Yorkshire terrier, a bitch and, again governed by instinct, it mated and eventually became the father of more of its own kind. This process occurred twice more in the next few years. Eventually the cells in the aging animal became damaged and unable to repair themselves adequately and the dog entered its final years. However, at the time of its death it was a great-grandparent and there were more than 40 other Yorkshire terriers that owed their existence directly to him.

This description of the birth, life and death of a living organism contains many distinctive parts but it is clear that not all of them are essential for what we call life. For example, a tree is a living system but it does not learn from experience and it cannot be trained to roll over. Many scientists have given definitions of what it is that constitutes life but they do not all agree. However, there are some common features in their definitions that we give here.

29.2 The Characteristics of Life

Our description of the birth, life and death of a Yorkshire terrier introduced some, but not all, of the characteristics of life. Here we give a set of characteristics that are certainly associated with all living entities.

29.2.1 *Reproduction*

Whatever definition is found for the state of being alive, any living mechanism will, at some time, die, so that it no longer satisfies that definition. The lifespans of living organisms vary enormously from a few days for the common fruit fly to nearly 200 years for some tortoises and then to an age of over 4,700 years for a presently-living Bristlecone Pine. But whatever the lifespan, be it long or short, the organism will eventually die. For this reason a necessary property of any living organism, on which all scientists agree, is that it must be able to reproduce itself in some way. The possible modes of reproduction are many and varied and can be as complex as the production of mammalian young, as for the Yorkshire terrier, or as primitive as the dispersion of spores by some plants, algae and fungi.

29.2.2 *Adaptation*

Over long periods of time, environmental conditions — temperature, humidity, type of vegetation, etc. — will change and these changes may be detrimental to the survival of the living organism. The changes in the organism that enable it to survive and flourish in the new environment can take place in a gradual way or sometimes in a more abrupt way through a process called mutation, which will be discussed more fully in Chapter 31.

29.2.3 *Regeneration and Growth*

Many living organisms are complex collections of cells which are organised to carry out different functions. Cells have a limited period of effective activity and so must be replaced for the organism to survive. Another characteristic is that the living organism must grow and develop from the time of its birth to some mature state. This requires the formation of cells other than just replacements for those that cease to operate.

29.2.4 *Metabolism*

Living organisms must be able to take in food in some form — non-living matter, dead animal or vegetable material — and to convert it either into energy or into proteins and nucleic acids from which new or replacement cells are constructed. The chemical reactions that occur in living cells and enable this to happen go under the collective name of *metabolism*.

29.2.5 *Response to Environmental Stimuli*

In order to optimise its ability to survive, a living organism must be able to react to stimuli from its environment. Such responses cover a wide spectrum of behaviour. Some unicellular organisms (that may not satisfy *all* the criteria for life on some definitions) contract when touched. A plant will turn its leaves towards the Sun to maximise its intake of solar energy. Through a complex set of reactions, which may involve many senses and analysis by the brain, an animal will either run to escape from impending danger or run towards a potential prey.

Some scientists would say that the above list of requirements to define life is inadequate while others might think that it is over-prescriptive. There are always borderline cases that are difficult to fit into either the category of the living or the inanimate. For example, viruses consist of genetic material (deoxyribonucleic acid, DNA, or ribonucleic acid, RNA — of which more in following chapters) wrapped up in a protective protein coat. If they invade a living cell they use the contents of that cell to replicate themselves. In the process the cell becomes damaged, or may even be killed, and the organism of which the cell is a part may suffer disease and even death as a result. Another life-like quality of viruses is that they adapt — indeed, they modify their structures so readily that it is difficult to develop antiviral therapies that are effective for any length of time. So viruses reproduce, given the right environment, and they do adapt, but they have none of the other attributes of life as defined here.

Chapter 30

Forms of Life

To those fortunate enough to live in fertile regions, with adequate and not excessive rainfall, the world teems with life. Everywhere there are trees, shrubs, flowers and grass. Birds fly in large flocks, cattle and sheep roam over rich pastures and fish can be seen in the clear waters of ponds, rivers and streams. Apart from this evident larger-scale life there are smaller life forms such as insects of various kinds and worms, which we can also easily see. At an even smaller scale, with a microscope, we find that pond water is chock-a-block with tiny microscopic life forms, as is the soil within which the insects live. However, even in areas of the world that are apparently almost barren, there is an abundance of life. A desert area will seem almost lifeless but after a brief period of rain it may become a carpet of brightly-coloured flowers; the life was always there but just waiting for the rain before revealing itself. The frozen Arctic and Antarctic regions may seem to be hostile environments for life but, apart from the obvious large inhabitants such as polar bears, seals and penguins, the seas are rich in fish and other sea creatures. The small marine animal, krill (Figure 30.1), which lives in these regions and feeds on microscopic phytoplankton, is itself a food source for many other creatures, including some species of whales. The estimated total mass of krill in the world, 500 million tonnes, is somewhat more than the total mass of humanity!

Faced with this abundance of different forms of life it is clearly not possible to exhaustively describe them all. However, on the basis of various criteria, scientists now divide life into three main *domains of*

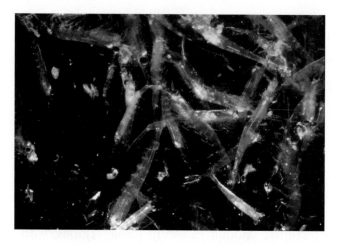

Figure 30.1   A swarm of krill (NOAA).

life. These are designated as *bacteria, eukaryota* and *archaea*, for each of which the main characteristics will now be described, and examples will be given of life forms in each of the categories.

30.1 Bacteria

These micro-organisms are unicellular, that is, they consist of a single cell. They are very small, typically a few microns long, and can take on many forms such as spheres, rods and spirals. A spherical bacterium, found in Arctic ice, is shown in Figure 30.2; one wonders how it got there and what it does there?

Most bacteria have a similar basic structure, illustrated in Figure 30.3, although there are some slight variations.

The components of the bacterium structure are as follows:

Cell wall This structure defines the shape of the bacterium and it consists of a mixture of proteins and polysaccharides, which are polymers of sugars, of which one example is cellulose. *Mycoplasma* is a type of bacterium that does not have a cell wall and hence has no definite or permanent shape. Some species of this genus of bacteria are harmful to human health — e.g. the one that causes pneumonia.

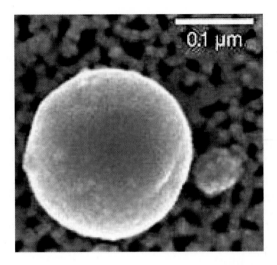

Figure 30.2 A bacterium found in Arctic ice (National Science Foundation).

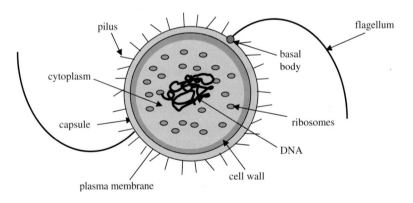

Figure 30.3 Representation of a bacterium.

Plasma membrane This is the layer through which nutrients pass into the cell and waste materials leave it. It consists of lipid materials, i.e. ones that are soluble in organic solvents and insoluble in water. Human body fat is one form of lipid.

Cytoplasm This is a jelly-like material within which reside the ribosomes and the genetic material DNA (deoxyribonucleic acid, Section 32.3).

Ribosomes There are many of these small organelles (components of an organism) in the bacterial cell giving it a granular appearance in an electron micrograph. Ribosomes are the locations where the proteins required by the cell for its survival and development are produced. Within them is a material called mRNA (messenger ribonucleic acid, Section 32.5) that controls which proteins are produced.

DNA (Deoxyribonucleic acid) DNA is the basic genetic material, the structure of which is a blueprint that defines every feature of an organism. It is the information contained in DNA that produces the mRNA needed by the organism to create the proteins it requires.

Pili (plural of pilus) These hair-like hollow structures enable the bacterium to adhere to another cell. They can be used to transfer DNA from one cell to another.

Flagellum and basal body The flagellum is a flail-like appendage that can freely rotate about the basal body, which is a universal joint, and so provide a mode of transport for the bacterium. A single bacterium may have one or several flagella (plural of flagellum).

There are many different metabolic pathways followed by bacteria, which can be categorised in two different ways. The first is by the source of energy they use to drive their nutritional needs and the second is the source of carbon from which the organic materials they require are produced. According to the energy sources they use, bacteria can be categorised as:

Phototrophs — using light as the energy source.

Lithotrophs — using the energy from reactions involving inorganic materials, e.g. ammonia (NH_3), hydrogen sulphide (H_2S) or iron (Fe) in its ferrous state, i.e. in a compound where it is bivalent (valence two).

Organotrophs — using the energy from reactions involving organic material, e.g. decaying, ex-living organisms.

The source of the carbon for making the structure of the bacterium is either *heterotrophic*, meaning that organic materials are the

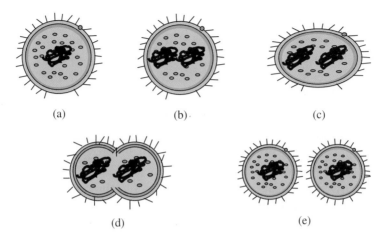

Figure 30.4 Fission of a bacterium: (a) A fully grown bacterium; (b) The DNA replicates; (c) The cell elongates; (d) The cell wall forms a neck; (e) Two daughter cells form.

source, or *autotrophic* where the carbon compounds are obtained by fixing atmospheric carbon dioxide (CO_2). Thus a bacterium described as *photoautotrophic* derives its energy from light and fixes carbon dioxide as its source of carbon.

The most common form of reproduction of bacteria is by cell division, illustrated in Figure 30.4. When the cell has grown to some critical size the DNA replicates, then a neck forms in the cell wall and, finally, the cell breaks into two, with each daughter cell containing a full complement of DNA and a share of the cytoplasm and other contents of the original cell. Each daughter cell will then grow until it reaches the size to divide again.

Bacteria are found in a wide variety of environments — almost everywhere except in the harshest environments. It is estimated that there are about ten times as many bacteria residing on and in a human body as there are cells that constitute the body itself, mostly residing on the skin and in the digestive tract. With a world population of bacteria estimated at about 5×10^{30} and with a mean volume of 10^{-18} m^3, if they were all lumped together they would occupy a cube with sides of length about 15 kilometres. Another comparison is that the mass of bacteria on Earth exceeds the mass of all humanity on Earth by a factor of between 1,000 and 10,000! Bacteria may be tiny but there are many of them.

It is generally believed that bacteria are harmful and that the world would be a better place without them. This is far from the truth. Some are dangerous; millions of people die each year from bacterial infections such as tuberculosis and cholera. However, through its immune system the body normally provides us with protection against many bacteria and when it does not do so we can call on antibiotics that deal with most of them. On the positive side, bacteria can be very useful and even necessary to other life forms. They are essential in recycling nutrients; when a tree dies, bacteria break down its substance so that it can be absorbed in the soil and be taken up by other plants. Bacteria, which are present in the roots of some plants, carry out the fixation of nitrogen from the atmosphere — they act as natural plant fertilisers.

30.2 Eukaryota

These are organisms for which the cells are of a complex form, where the genetic material is contained within a nucleus bounded by a membrane. Within this domain there are four *kingdoms*, sub-groups within the domain, of connected distinctive organisms.

30.2.1 *Protista*

These are single-celled organisms plus some of their very simple multi-celled relatives. Single-celled protists include amoebas and diatoms while slime moulds and various types of algae are examples of the multi-celled variety. An example of this kind of organism is the red algae, shown in Figure 30.5. Red algae are often organised into a plant-like form and they are harvested and used as food in some societies.

Since they are so simple and can consist of a single cell, protists do have some properties in common with bacteria. The primary difference is that protists are *eukaryotic*, which means that they have a

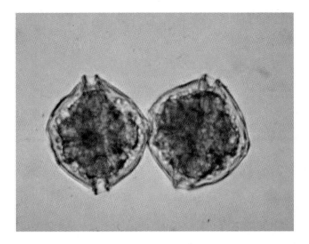

Figure 30.5   Red algae (Woods Hole Oceanographic Institution).

nucleus with DNA bound within a membrane. By contrast bacteria are *prokaryotes*, where the DNA is in a single strand within the cytoplasm.

Protists obtain their nutrition in a number of different ways. Where they possess flagella, similar to those described for bacteria, they can filter feed, where the flagella find the food source and bring it to the membrane to be absorbed. Some protists can wrap themselves round bacteria and absorb them directly through their membranes. Other types of metabolism resemble those found in bacteria; they can use a phototrophic process in which the carbon source is either organic material or carbon dioxide or an organotrophic process in which organic materials are the carbon source.

Some protists reproduce asexually, as has been described for bacteria, while others use sexual reproduction where parts of the DNA from two cells combine to form a new DNA strand with random components from the two contributors.

30.2.2 *Fungi*

These are a very common type of organism, mostly multi-cellular, of which more than one million different varieties are known to exist.

Figure 30.6   A woodland fungus (T. Rhese, Duke University).

Mushrooms, toadstools, various moulds and yeasts are typical examples of fungi. They digest their food externally and then absorb the digested nutrients into their cells; thus many mushrooms are found on rotting fallen trees, which are their source of food. A typical fungus, as may be found in woodland, is shown in Figure 30.6.

Reproduction of fungi can be asexual, by the dispersal of spores that bear exact copies of the parent fungus' genetic structure, or by a sexual process called *meiosis*, which gives a mixture of the genetic structure of two separate spores and hence gives variation within the species. The process and advantage of mixing genetic material is described in greater detail in Section 32.6.

Mushrooms are a food source, a tasty but not a very nutritious one, and yeasts, another form of fungi, are used both for making bread and fermenting alcoholic beverages.

30.2.3 *Plantae*

These multi-cellular organisms include all land plants such as flowering plants, bushes, trees and ferns. There are about a quarter of a million known members of this kingdom. Almost all plants are green as a consequence of the pigment chlorophyll that they contain. This pigment uses the Sun's energy, by a process known as photosynthesis, to fuel the manufacture of food such as starch, cellulose and other carbohydrates that constitute the structure of the plant. The process for forming cellulose is described in words as:

$$\text{water} + \text{carbon dioxide} \xrightarrow{\text{Light energy}} \text{Glucose} + \text{oxygen}$$

or, as a formal chemical equation:

$$6H_2O + 6CO_2 \xrightarrow{\text{Light energy}} C_6H_{12}O_6 + 6O_2$$

Glucose is a sugar that in the form of a polymer, a chemically-bound long string of glucose molecules, constitutes cellulose. Cellulose is an extremely strong material from which the stems of plants are formed and which also enables plants to grow large structures like tree trunks. A Douglas fir tree is a fine example of such a structure (Figure 30.7).

Plants can reproduce either asexually or sexually. Asexual reproduction can occur in a number of ways. In the case of strawberry plants, for example, underground runners from the main plant become the basis for a new plant to grow. The bulbs of flowering plants divide and each section of the divided bulb is capable of developing into a full-sized bulb to produce a new flower. Leaf or stalk cuttings taken from many planets will develop roots and then grow into mature plants. Finally — and commercially important — a cutting taken from an apple tree and grafted onto a root stock ensures that the variety of apple is faithfully reproduced.

For sexual reproduction both male and female organs are required and often both are contained in the same plant. A typical bisexual plant, in the form of a flower, is illustrated in Figure 30.8. The male structures are the filament and the anther. The anther produces and contains pollen, which contains the male genetic contribution. The female part of the system consists of the stigma, style and ovary. Pollen is transferred from plant to plant by the wind, insects, birds or small mammals; the end of the stigma is sticky and so traps the pollen. Once the pollen grain is captured by the stamen then it germinates a

Figure 30.7   Douglas fir trees.

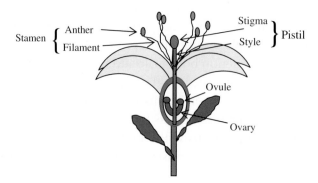

Figure 30.8   A schematic flower showing the major components of the reproductive system.

pollen tube that penetrates through the stigma, style, ovary and ovules so that two sperm cells that follow can reach the eggs, which produce both the fruit, or other seed-storage entity, and the seed itself.

30.2.4 *Animalia*

This is the kingdom to which humanity belongs. Unlike plants, which can manufacture food from non-living material, animals can only ingest the products of life — protista, fungi, plantae or other animals. Animals are also the only life form that has two distinct types of tissue — muscle tissue and nervous tissue.

There are a bewildering number of different classes in the animalia kingdom with members as diverse as *protozoa* (Greek for 'first animal') to ourselves, *homo sapiens*. The protozoa are simple single-celled creatures with flagella, which are similar to protists in many of their characteristics but qualify as animalia because they breathe, move and reproduce like multi-celled animals. A general classification that can be made in the animal kingdom is between *invertebrates*, those without backbones, and *vertebrates*, those with backbones. Within these general groups there are several different classes of creatures but we shall just describe a few types of creature in each of the invertebrate and vertebrate groups to illustrate the range of life forms and lifestyles in each group.

30.2.4.1 *Invertebrates*

Earthworms The general form of an earthworm is familiar to most people, in particular to gardeners, and is shown in Figure 30.9. Earthworms have a very simple blood circulatory system without a heart in the conventional sense but with a number of regions within the blood vessel where contractions act as pumps to keep the blood moving. This blood passes through a region close to the wall of the gut where gasses and nutrients can be exchanged. Food is taken in by suction though the mouth and is temporarily stored in a crop before being passed into a gizzard where a combination of sand and muscular contractions turns the food into fine mulch. This then passes

Figure 30.9   An earthworm.

through the intestine which absorbs the useful nutrients, with the waste being extruded through the anus at the rear of the creature.

Earthworms absorb oxygen and release carbon dioxide, the waste product of their metabolism, through their skins; for this to happen effectively their skins have to be damp but not too wet. In very wet conditions they may come to the surface to avoid saturated underground conditions.

Earthworms are hermaphrodites, which is to say that each of them has testes to produce sperm and ovaries to produce eggs — the egg-sac can be seen as a bulge in Figure 30.9. Normally, they do not fertilize their eggs with their own sperm but exchange with another earthworm, each providing sperm to the other's eggs. Earthworms can live up to 20 years but the average lifetime is in the range six to eight years.

Octopus The octopus (Figure 30.10) is a member of a class of creatures called *cephalopods*, which include squid and cuttlefish. They are

Figure 30.10 An octopus.

entirely soft-bodied, except for a beak, similar to that of a parrot. They have three separate hearts, two pumping blood through their gills where dissolved oxygen is extracted from the water in which they live, while the third drives blood around the animal's body to take the oxygen to where it is required and to remove carbon dioxide, the product of energy generation in the regions of its muscles.

In reproduction one of the eight arms of the male is used to insert packets of sperm into the female's body. Usually the male dies shortly after mating and the female often does not survive very long after laying about 200,000 eggs and looking after them until the young hatch out.

An octopus has several unusual physiological characteristics. It has two novel defence mechanisms; one is to eject clouds of black ink to cloak its escape and the other is produced by special skin cells that can change colour to match that of the surroundings and hence act as effective camouflage. Another unusual feature of the octopus is the keenness of its eyesight, better than that of humans in many respects.

Crabs The most obvious feature of a crab (Figure 30.11), in common with other crustaceans such as the lobster, is the presence of an external skeleton (exoskeleton), which gives rigidity to the body and protects a fairly primitive circulatory and nervous system.

Figure 30.11   A crab.

Because of its exoskeleton the growth of a crab cannot take place in a smooth continuous way. To grow the crab must first shed its shell then, in an unprotected state, undergo a period of growth and finally re-grow its exoskeleton.

The body of a crab is in three parts — head, thorax and abdomen. The head is equipped with two compound eyes, like those of insects, and antennae that can ascertain the external temperature and humidity and also assist in detecting food. Crabs possess gills that can extract dissolved oxygen from water but out of water it can exchange gasses with the outside world from all parts of its exposed body surface. There are separate male and female crabs and reproduction takes place by the transfer of sperm by legs of the male to eggs attached to legs of the female.

The front legs of the crab are equipped with two large pincer-like claws that are used for holding food and transferring it to the creature's mouth and also to tear mouth-sized chunks from the carcass being eaten. Crabs sometimes hunt for food but often they are scavengers, picking up the unwanted morsels left by other predators.

The housefly The final example of an invertebrate is a well-known example of an insect, the housefly (Figure 30.12), which possesses the characteristic of flight. They have two wings but some other insects, e.g. dragonflies, have four.

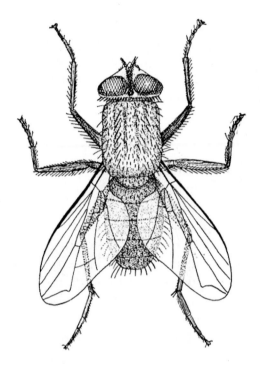

Figure 30.12 The common housefly.

Like all insects, houseflies have six legs with a body divided into three segments, head, thorax and abdomen. They hatch out of an egg as a white maggot that feeds, usually on decaying matter, until they are a few millimetres long. They then transform into a brownish pupa from which the adult housefly emerges. Long ago in older civilisations, and also more recently, it has been observed that wounds infected with maggots, usually in a wartime situation, would heal in situations where it was expected that a limb would need to be amputated. The maggots eat only dead, gangrenous flesh and prevent the corruption from spreading from morbid to healthy flesh. During World War II, before the introduction of penicillin, maggot therapy was used by the American army and it is sometimes used at the present time. Houseflies have a short lifetime, of order three weeks, and within a couple of days after emerging from the pupa female flies are ready to mate. The male injects sperm into the female, who stores the sperm that can then be used for fertilising several sets of eggs.

The primary diet of flies is decaying food that they ingest by first liquefying it by covering it with saliva and then sucking in the resultant fluid. They are generally regarded as a pest since they spread disease, although, in nature, they play the role of a scavenger that cleans up the environment to some extent.

30.2.4.2 *Vertebrates*

There are some obvious limitations to the size of an invertebrate creature. Without a rigid supporting structure, at a certain size gravity would begin to flatten the creature, so as to provide enough area to lessen the pressure on the supporting surface. This is similar to having a car with under-pressure tyres that flatten to support the weight of the car. If the supporting structure is to be external, as in the case of the crab, then for growth to occur the shell must be discarded from time to time, at a tremendous cost to the creature of growing a new shell, especially if the shell is large and thick.

The advantage of a backbone and accompanying skeletal structure is that a much larger range of sizes can be accommodated, from the smallest bat with a mass of about 2 grams to an elephant with a mass of several tonnes. However, even with a skeleton there are limits to the mass of a land-dwelling animal. If we start with a mouse and scale it up to the size of an elephant then it would be unable to support its own weight. Its mass would increase as the cube of the linear dimension but the cross section of the legs, which has to support the weight, would only increase as the square of the linear dimension. To enable our expanded mouse to remain upright the thickness and sturdiness of its legs would have to be greatly increased. This is why a spider can have thin spindly legs in relation to its body size while an elephant's legs are squat and thick. Mice and elephants, and many animals of intermediate size, including mankind, are *mammals*, which have as one of their common characteristics that they give birth to live young (with the single exception of the platypus that lays eggs) and that infants are fed with mother's milk. There are some mammals that inhabit the oceans, e.g. whales and dolphins, and in the case of whales, some species of which can have a mass of over 100 tonnes; the sea is the only place they could live since no conceivable functional legs could support and transport an animal of that mass on land. It needs the buoyancy of water to support its weight.

Against that background we now describe a sample of four types of vertebrate.

Fish Fish are cold-blooded vertebrates that occupy most large bodies of water, either fresh or salt, from the tropics to the Polar Regions. In very cold environments their blood may contain an antifreeze component that stops the blood temperature from falling too close to the freezing point, so becoming viscous and less able to flow easily. They vary enormously in size — the smallest being less than 1 centimetre in length and the largest, the Whale Shark with a mass up to 36 tonnes. Most fish are of a streamlined shape, so that they move quickly through water with the minimum expenditure of energy. They breathe through gills, situated on the two sides of the fish at the back of the head, through which water is drawn to pass through a system of fine capillaries in which oxygen is extracted from the water and exhaled carbon dioxide is dissolved in it.

Fish can be herbivores, carnivores, omnivores or scavengers. The largest fish, the Whale Shark, is a filter feeder, meaning that it passes huge quantities of water through its mouth and filters out small food particles suspended in the water — plankton, krill, small fish and squid. By contrast the smaller Great White Shark — still a monster of up to 5 tonnes — is a ferocious predator of smaller fish and a danger to human bathers in some areas of the world.

The most common method of reproduction is known as spawning. The female fish deposits her eggs in some suitable environment and then the male fish swims over the region of the deposited eggs releasing its sperm in a great cloud that makes its way to the eggs to fertilise them. The number of eggs deposited by each fish is extremely large but the survival rate of the fry (newly hatched fish) is tiny, so fish numbers do not expand rapidly. There are some fish, such as salmon, that live most of their lives in the sea but make their way into fresh-water rivers, to the places where they were born, to spawn. The great

Figure 30.13   The Atlantic cod — an important food source for humans.

Figure 30.14   The Siamese crocodile.

majority of salmon die after spawning — they have then achieved their biological purpose and are surplus to the requirements of the species.

Fish, e.g. cod (Figure 30.13), are an important food source for humans but over-fishing of some species is leading to the possibility of their extinction.

Crocodiles Crocodiles are large aquatic reptiles, up to nearly 5 metres long and weighing more than one tonne (Figure 30.14), which live in many parts of Africa, Asia, America and Australia. Their bodies are covered with protective scales, giving them a prehistoric appearance although they are highly developed animals. Unlike other reptiles, they have in common with mammals a four-chambered heart and a cerebral cortex, the part of a mammal's brain that is responsible for the most highly-developed sensory activity — awareness, consciousness and thought, for example — but the extent to which a crocodile can engage in these activities is unknown.

The crocodile is a fearsome predator, able to move swiftly in water and over land for short distances. Its mode of attack is to seize the prey in its jaws, which are capable of applying a pressure some ten times that of a Rottweiler dog. Its numerous and sharp teeth are ideal for holding on to its prey and tearing its flesh once it begins to eat. It often kills its prey by holding it under water until it drowns and then it may keep the dead animal under water for some time until decay softens its flesh and makes it easier to consume.

Their lifetimes are similar to those of man — a typical lifetime is 70 years but some individuals may live to 100 years or more. Reproduction is by the male inserting sperm into the female who eventually lays in the region of 40 to 60 eggs in a prepared nest, which must neither be submerged in water — which would drown the young — nor too dry — the eggs must stay moist during incubation. After an incubating period of 80 days the young emerge. The embryos do not have separate chromosomes that define the sex of the newborn crocodile. Sex is determined by the temperature during the incubation period — in the region of 31.6° C males are born but at higher or lower temperatures the eggs produce females.

Peregrine falcon Birds occur in a great range of sizes and lifestyles from the *bee hummingbird* with a weight of 1.6 grams, which hovers as it feeds on nectar from plants, to the *ostrich*, nearly 3 metres tall with a weight of more than 150 kilograms, which is flightless. The bird we have chosen to represent this class of animalia is the *peregrine falcon*, a bird of prey that inhabits many regions of the world. Figure 30.15 is a reproduction of a plate from Audubon's *Birds of America* (published between 1827 and 1839), showing this magnificent bird.

Figure 30.15   The peregrine falcon (Audubon).

In common with other birds, the peregrine falcon is feathered, warm-blooded, bipedal and lays eggs to reproduce. Typically it is 45 centimetres long with a wingspan of 1 metre and its main food is smaller birds and sometimes small mammals. It mainly hunts at dawn and dusk and it seeks its prey either from some high perch or from the air. It catches its prey by folding back its wings and going into a high velocity dive, estimated to sometimes reach over 300 kilometres per hour.

These birds have been used for falconry for more than 3,000 years. They are sometimes employed on airfields to scare away other birds that could endanger planes taking off or landing by being sucked into their engines.

Primates The primate class of animalia contains a large range of creatures — including lemurs, monkeys, gorillas, chimpanzees (Figure 30.16) and humans. Primates have forward-looking eyes, giving good distance judgement through stereoscopic vision, and a bony ridge above the eyes, which is less prominent in man than in other primates. However, the most important common characteristic is five digits on each limb, with nails, pads on the fingertips and, for most primates, an opposable thumb that makes the hands able to manipulate objects in a very subtle way and to construct and use tools. When this manual dexterity is combined with the large brain size of man it

Figure 30.16 Man's closest relative — the chimpanzee (National Library of Medicine).

creates an animal with a level of attainment of a different order of ability to that of any other animal. In round figures the capacity of the human brain is about 1,200 cubic centimetres. For gorillas, chimpanzees and orang-utans the capacities are 470, 400 and 400 cubic centimetres, respectively. Although some animals, such as the elephant and the whale, have much larger brains than man they do not have the other physical attributes that have enabled man to use his brain capacity so effectively.

Primates form strong social groups, usually with one dominant male as leader of the group, and they display social behaviour such as the sharing of food, the nursing of children of other parents and mutual grooming. There is also a tendency for young females, and sometimes males, to leave the groups in which they were born to join other groups. This is also a pattern of behaviour of tribes in New Guinea where it is commonplace that a man from one tribe will take as a wife a female from another; this is sensible as it reduces the risk of potentially harmful inbreeding within a small group. Humans are omnivores, as are chimpanzees and some other primate species. However, most species of primates are herbivores, although they often supplement their diet with insects.

30.3 Archaea

Before 1977 the above account of living species as either bacteria or eukaryota would have been considered complete. It would have been noted that there were some bacteria that had adapted to living in very harsh conditions of high temperature and salinity; such organisms were found in various pools of the hot springs of Yellowstone Park and in some other environments. Then in 1977 these 'adapted bacteria' were subjected to analysis by Carl Woese (b. 1928) and his colleagues from the University of Illinois and were found to be genetically and structurally distinct from the bacteria that otherwise they so closely resembled. They look like bacteria under a microscope and, like bacteria, they can take on a great variety of forms, for example, a filament-like structure as shown in Figure 30.17 — but they are not bacteria. They also take up shapes that are not found for bacteria, for example, flat triangles or oblongs. This gave rise to a new domain of life — *archaea* — which all have the characteristic that they can exist at very high temperatures and in chemical environments that would be toxic to most life forms. In addition they do not require oxygen to survive — a necessary condition to be able to live before atmospheric oxygen was formed.

Archaea are found in many locations with extreme conditions of temperature and salinity or alkalinity. They are found at the bottom of the ocean near volcanic vents that heat the water to over 100° C. They occur in the digestive tracts of cows, termites and some sea creatures, where they produce methane. In fact, it is argued by some climatologists that the methane emitted by cows is a serious contributor to global warming and, if so, it is archaea that are the ultimate

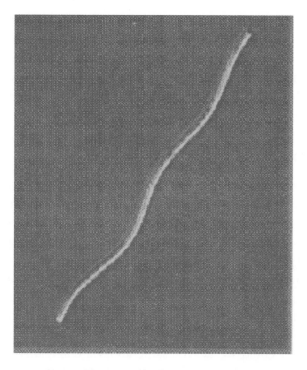

Figure 30.17 A filament form of archaea.

culprits! Other places where archaea are found are in marsh mud and even in petroleum deposits. All these different archaea, distinguished by their different habitats and 'lifestyle', are clearly genetically related and could be adaptations from a single source.

An essential characteristic of any life form is that there must be some metabolic process that will provide it with the material to maintain its structure and also that there must be some reproductive process that will create replacements for those entities that die. The metabolic and reproductive processes for archaea are those noted for bacteria plus other anaerobic (oxygen-free) metabolic processes that use the presence of sulphur or methane, present in sulphur springs, volcanoes or mineral deposits.

Some features of archaea seem to resemble those of bacteria and eukaryota more than those last two domains resemble each other, leading to the postulate that archaea were perhaps the first life forms; they could have existed in the hostile environments offered by the early Earth when other life on Earth would have been impossible. The recognition that life can exist in harsh environments has encouraged the hope that some life forms may be found in extreme conditions on other solar-system bodies — Mars, Europa or even Venus.

30.4 An Overview

This survey has revealed the tremendous range of complexity of living organisms and their various physical characteristics but it has not covered other characteristics that cannot be described in physical terms. The most notable of these is *consciousness*, a human quality that describes a whole gamut of mental activity. We are aware of our own existence and our place within our environment. We have thoughts relating to past activity and future planned activities. We have feelings of one sort or another — sorrow, fear and happiness, for example. Nobody would ascribe the property of consciousness to a bacterium or a tree, but could we say the same about a chimpanzee? An ant, a communal creature that acts in unison with others, is a member of the kingdom of animalia. A soldier ant will sacrifice its life for the good of its colony, but could one associate such a sacrifice with courage based on reason and sentiment– or is the ant just programmed like a unicellular organism that recoils to touch? It is doubtful that one could ascribe consciousness to an ant although one could find parallels between some of its activities and those of man.

The fact of life itself is so immensely significant that it can be said with confidence that the difference between a man and a bacterium is trivial compared with the difference between a bacterium and any non-living object. But how did this life begin? As stated at the very beginning of this chapter, this is a question without an answer. But at least we can discuss the nature of the problem — as we shall do in Chapter 33 — and also describe how it is that, once a primitive, simple living organism has been produced, more complex life forms, up to and including man, can evolve.

Chapter 31

Nineteenth Century Genetics — The Survival of the Fittest

31.1 Mendel and His Peas

'Oh, he does look like his father.' How many times has that been said by a simpering friend or relative looking at a baby in a pram? In fact, very few babies have established features that clearly resemble one or other parent but, once they reach the toddler stage, parental characteristics can often be seen. Some centuries ago the common view was that offspring tended to show characteristics that were a blend of those of the parents — although it must have been observed that the colour of the eyes of the offspring of a blue-eyed father and a brown-eyed mother was not a muddy blue.

The scientific field that is concerned with the way that characteristics are passed from one generation to the next is called *genetics*. The person who began this science, often referred to as the 'father of genetics', was an Austrian Augustinian priest, Gregor Mendel (Figure 31.1). Between 1856 and 1863 he carried out experiments involving the growth of common pea plants over many generations. He noted that certain characteristics of the plants he bred were strictly one of two alternatives and never a mixture of the alternatives. For example, the pea flowers were either white or purple and in no situations were intermediate colours produced, i.e. a very pale purple. In total, Mendel observed seven different properties of the peas that could be one of

Figure 31.1   Gregor Johann Mendel (1822–1884).

two kinds, but never a blended mixture of the two kinds. These properties were:

(a)   The flower colour is white or purple.
(b)   The seed (pea) colour is yellow or green.
(c)   The stem length is long or short.
(d)   The flower position is either at the end of the stem or in some intermediate position.
(e)   The pod shape is flattish or inflated.
(f)   The seed is smooth and round or wrinkled.
(g)   The pod colour is yellow or green.

The pea flowers have both male and female sexual organs so that fertilisation can be either by *cross-pollination*, where one plant is pollinated by another, or by *self-pollination*, where the plant pollinates itself.

We can illustrate Mendel's experiments by considering the colour of peas — either yellow or green. Mendel first began with two peas of different varieties, one of which produced only yellow peas and the other of which produced only green peas. He then grew plants from the two varieties and artificially cross-pollinated so that the female sexual organ (*pistil*) of each plant was pollinated with pollen from the male sexual organ (*stamen*) of the other plant. The first generation plants, conventionally referred to as f1, gave only yellow peas and not, as the common view would have it, some form of yellowish green. This, and subsequent, stages of the experiment, in which only self-pollination was used, are illustrated in Figure 31.2.

Next the yellow peas from the f1 generation were self-pollinated and the result of this, in the f2 generation, is that three quarters of the peas were yellow and one quarter were green. In the third, f3, generation it was found that one third of the yellow peas only produced yellow peas, the other two thirds producing three quarters yellow and one quarter green. Finally, the progeny of the green pea from the f2 generation were all green. This pattern of results was found for all the other characteristics examined, and Mendel's great contribution was to interpret what these results implied for the inheritance of physical characteristics. We now describe Mendel's interpretation in terms of the pea colour.

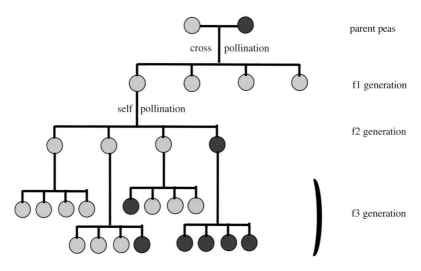

Figure 31.2 A representation of Mendel's experimental results for the colour of peas.

The first conclusion is that the colour of the pea is controlled by entities within the plant, entities that we now know as *genes*. For each gene there are two different forms, called *alleles*, and each plant contains two alleles that can either be of similar form (in which case, for this characteristic, the plant is *homozygous*) or of different form (in which case, for this characteristic, the plant is *heterozygous*). The original parent peas were both homozygous, one having the alleles both yellow (Y) and the other having the alleles both green (G). The f1 generation of plants inherits one allele from each parent so that in this case all the f1 generation had the alleles Y + G. Now we must address the question of why *all* the f1 generation were yellow. The answer is that one allele is *dominant* and the other is *recessive*. In this case the Y allele is dominant and hence all the heterozygous plants (Y + G) of the f1 generation were yellow. In outward appearance they were identical to the yellow parent plant but hidden from sight was an important difference — the recessive green gene.

We now come to the f2 generation, produced by self-pollination of the f1 peas. There were no longer two parents but the genetic composition was passed on by pollen representing the male contribution, and the female pistil, each of which contains the gene-pair Y + G. Which allele was passed on to the offspring was quite random, so there were four possible combinations in the offspring:

$$Y + Y \quad Y + G \quad G + Y \text{ (equivalent to } Y + G) \text{ and } G + G$$

and on average each allele combination was passed on to one quarter of the progeny. However, since Y is dominant three quarters of the offspring were yellow and only one quarter was green, corresponding to G + G. The only way that the recessive character could appear in the plant, being green in this case, is if the plant was homozygous with both alleles of the recessive G kind. In Figure 31.3 we now repeat Figure 31.2 but add the genetic makeup (genotype) of each plant. You should be able to follow the figure through to generation f3 using the principles described above.

This basic structure of genes, with two alleles of which one is dominant, explains Mendel's results and many other observations about inheritance. In most European counties there are some people

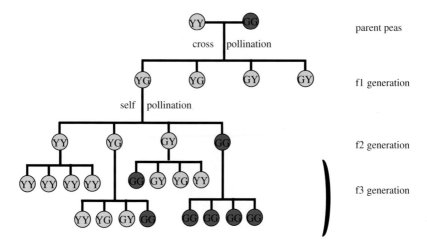

Figure 31.3   Figure 31.2 with added genotypes.

with blue eyes; for the gene controlling eye colour the alleles correspond to brown and blue, with brown dominant. People of African and far-Eastern origin have brown eyes since the blue allele is completely absent in those populations; where the proportion of one allele is 100%, so the partner gene is absent, then this allele is said to be *fixed* in the population.

Another result that came out of Mendel's experiments is that the outcomes for each of the seven characteristics were completely independent of each other. The physical characteristics were controlled by different genes and random selection prevailed for each of them separately.

Mendel's work was purely empirical and at the time it was done there was no possibility of finding the basic physiological mechanism that controlled its operation. However, at the same time that Mendel was carrying out his seminal work there were observations being made that related to what he was doing, although the connection was never made at the time.

31.2 The Discovery of the Chromosome

Working on a different aspect of genetics, and completely unaware of Mendel's work, a German biologist Walther Flemming (Figure 31.4)

Figure 31.4   Walther Flemming (1843–1905).

made important observations in the 1870s on the process of cell division, which he called *mitosis*.

By the use of synthetic aniline dyes, which the German chemical industry was producing in great numbers in those days, Flemming discovered that there was a structure within the cells that strongly absorbed the dyes and hence was easily visible within the cell under microscopic examination. He gave the structure the name *chromatin* and found that they were correlated with fine threadlike structures, later called *chromosomes*.

Flemming studied the process of cell division and established the principle that all cells were derived from identical cells, but he did not discover the way that the chromatin in the daughter cells from division was derived from the parent cell. It was only much later that the nature of chromatin and the threadlike structures and the way that mitosis operates was completely understood.

31.3 Lamarck's Evolutionary Ideas

The propagation of peas just produces more peas similar to those that preceded them, but quite early on there arose the idea of

Figure 31.5   Jean-Baptiste Lamarck (1744–1829).

evolution — the idea that a living species could evolve and change its characteristics. The distinction of being the first to put forward this idea belongs to a Frenchman, Jean-Baptiste Lamarck (Figure 31.5), whose first career was as a soldier but who later became interested in natural history.

Lemarck had the idea that during its lifetime a living organism would be modified by its environment or lifestyle and any such changes would be passed on to the progeny of that individual. An example of this is that a blacksmith, by virtue of his work, would become very muscular and that this characteristic would be passed on to his children. It was an interesting idea, if not one held today, and really brought into focus the concept of evolution, the way that species can change over long periods of time.

Lamarck's ideas actually had a revival in the Soviet Union in the 1940s and 1950s, during the time that Joseph Stalin was the leader of that country. An agronomist, Trofim Lysenko, claimed that wheat could be modified by subjecting it to low temperatures to produce strains that would be able to grow early in the year so that extra crops could be raised. Lysenko's ideas, which differed from the prevailing evolutionary theory in the rest of the world, fitted well with the general philosophy of the Soviet Union that regarded conventional genetics as a 'bourgeois science' and he was strongly supported by Stalin. This attitude may also have been encouraged by the previous rise of the National Socialist regime in Germany, which, while it lasted, had been the ideological enemy of the Soviet Union. The National Socialists promoted the idea of German or Nordic peoples as a 'master race' and were keen on the idea of eugenics, which is based on selective breeding of humans to produce desirable characteristics. In fact, selective breeding *is* carried out quite widely in the world, but is only applied to such things as racehorses, cattle and plant life.

31.4 Darwin's Evolutionary Ideas

The next, and most important, figure in the field of evolution studies is Charles Darwin (Figure 31.6), one of the giants of British science in comparatively modern times. Darwin was a contemporary of Mendel but their work did not really overlap and each was unaware of

Figure 31.6   Charles Darwin (1809–1882).

the work of the other. Darwin was the son of an eminent doctor and started as a medical student at Edinburgh University. He did not take to this — he was repelled by the butchery that passed as surgery in those days — and later began a theology course. While at Edinburgh he joined a student group interested in natural history, the Plinian Society, as a result of which he learnt of the work of Lamarck, which received wide acceptance in academic circles at that time. Later, at the insistence of his father, he moved to Cambridge University to complete studies that would equip him for the life of a clergyman. During his time at Cambridge he came under the influence of the Reverend John Stevens Henslow, Professor of Botany, and he attended Henslow's course on natural history. Darwin eventually graduated from Cambridge and was set to take Holy Orders and to become a clergyman. Then there occurred the critical event that shaped the rest of his life. Henslow recommended Darwin to be the naturalist on the voyage of HMS *Beagle*, a ship commanded by Captain Robert Fitzroy, which was about to begin a planned two year journey to map the coast of South America. In fact, the journey took five years.

During the voyage of the *Beagle*, Darwin made many and varied observations on both living organisms and on fossils and also found evidence of large geological upheavals such as the presence of sea-shells at high altitudes in the Andes. He took a particular interest in the fauna of the Galapagos Islands, where he noted that various species of birds and tortoises had different characteristics on different islands.

The *Beagle* returned home in 1836 by which time, by virtue of the specimens and notes he had sent back to Cambridge, Darwin was well known in scientific circles. Darwin had at this stage collected virtually all the data that was needed to make his great scientific contribution but, although he had mentally formulated his theory of evolution, he proceeded cautiously because he knew that a flawed presentation of his ideas would be savaged by both the scientific and religious establishments. In 1856 a friend of Darwin, Charles Lyell, read a paper by another naturalist, Alfred Russel Wallace (Figure 31.7), that seemed to present ideas very similar to those of Darwin. Alarmed that, after the long period of carefully preparing a presentation of his ideas, those

Figure 31.7   Alfred Russel Wallace (1823–1913).

ideas could be presented by someone else, Darwin started on a book called *Natural Selection*. However, in 1858 Wallace sent him a paper entitled *On the Tendency of Varieties to Depart Indefinitely from the Original Type*, which described the basis of natural selection and Darwin, although shocked by this pre-emption of his own ideas, generously arranged for Wallace's paper to be published. Eventually, in 1859, Darwin presented his own theory of natural selection in a book the full title of which is *The Origin of Species by Means of Natural Selection or the Preservation of Favoured Races in the Struggle for Life*, but is usually referred to as *Origin of Species*. It must be said that Darwin and Wallace became friends, communicated cordially with each other throughout this period and each appreciated the contribution of the other in a true spirit of scientific enquiry. Such generosity of spirit is, alas, less common today in the scientific community, where the struggle for scientific survival often involves the procurement of scarce financial resources.

So what is this principle of natural selection that was described in *Origin of Species?* Basically what it says is that within any species there are variations of characteristics so that, for example, there is variation in the height of men and in the coloration of insects. Sometimes in the prevailing conditions, which could change either gradually or suddenly, some variations of a particular characteristic could become more favourable to survival than others. An example that is sometimes quoted is that, if climatic change produced trees with leaves at a higher level, then those leaf-eating individuals with longer necks would be better equipped to survive. Eventually, within the species, the individuals with longer necks, whose progeny would take after their parents, would become dominant. Over a long period of time, after several changes of characteristics a new species could be produced, substantially different from the original source species. Taken to its limit, operating over long periods of time, an evolutionary pathway can be traced from a simple single-celled organism through to man. Starting with the same single celled organism, by different pathways one can end up with plants, insects and various other life forms including humans. A simplified evolutionary tree is shown in Figure 31.8.

Despite the care with which Darwin prepared his exposition of natural selection his book released a fierce storm of argument. In particular, the idea that primates and man were cousins in an evolutionary sense was violently opposed by the clergy, and others who were fundamentalist in their biblical beliefs. Darwin was subjected to both abuse and ridicule and cartoons appeared in periodicals showing apes with Darwin's head superimposed. Lest it be thought that this was just the ignorance of Victorian England, as late as 1925, in the famous 'Monkey Trial', the American state of Tennessee prosecuted a schoolteacher, John Scopes, for teaching Darwin's theory of evolution. Even today, a number of American states insist that creationist theory, the biblical version of the way that the world and its living contents were created about 6,000 years ago, should be given equal weight with Darwin's theory when taught in schools.[a]

The great showdown with respect to Darwin's ideas came in a debate at the British Association for the Advancement of Science

---

[a] See for example, Slevin, P., "Battle on teaching evolution sharpens", *Washington Post*, 14 March 2005, p. A01.

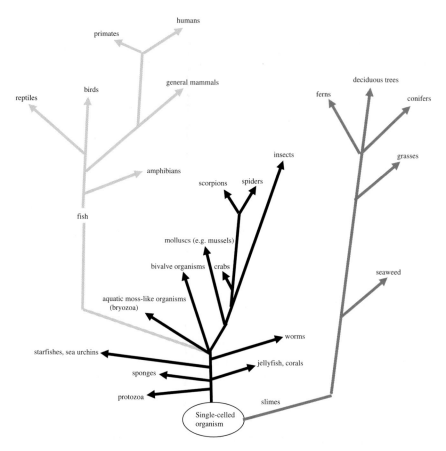

Figure 31.8   A schematic evolutionary tree. Representative organisms are given in most places rather than group names that would be unfamiliar to most readers.

meeting in Oxford in June 1860. Darwin, who suffered from chronic poor health, was not present but the main protagonists were Samuel Wilberforce (1805–1873), Bishop of Oxford and a previous President of the BAAS, who opposed Darwin's ideas, and Thomas Huxley (1825–1895), a strong supporter of Darwin. There was no outright winner of the debate itself — Huxley had the better arguments but Wilberforce had many supporters in the audience — although the debate exposed Darwin's theory to a wider public and so helped to promote it.

Darwin's theory was based on the knowledge of variability in the characteristics of living organisms that would confer advantages or disadvantages for survival under various conditions. We now know that the variations depend on the gene structure expressed in the organism's DNA although the variation may also be dependent on the conditions in which the organism lives. A gene that controls eye colour will give blue eyes or brown but a gene controlling size will also depend on the quantity and type of food available to the organism. During World War II, with the introduction of stringent rationing of food in the UK, which gave a balanced diet, the health of the nation improved and children were actually taller than their pre-war counterparts. Again, Japanese Americans have tended to be taller than their kin who remained in Japan and this reflects the differences in American and Japanese dietary patterns rather than a difference in their genes.

We have used the terms 'DNA', 'genes' and 'chromosomes' in the knowledge that these are now common terms in the vocabulary of most well-read individuals. In the following chapter we shall give a scientific framework for these terms and also explain how the genetic changes occur which enable the principle of 'survival of the fittest' to operate.

31.5 A Mathematical Illustration of Survival of the Fittest

Before delving into the intricacies of the scientific description of DNA and related phenomena we can illustrate the 'survival of the fittest' principle by a mathematical approach with the following example. Let us consider a gene that controls an important characteristic of the organism with alleles represented by A and a, with A being dominant. Then the possible genotypes are AA, Aa $(= aA)$ and aa. Since A is dominant then the organism will have the characteristic corresponding to allele A unless it is of genotype aa. Now let us suppose that there is a slight disadvantage in having the characteristic corresponding to A such that while all aa genotypes survive to maturity, i.e. to an age for breeding, only 99% of the other genotypes do so. If we begin with a generation in which the alleles A and a are equal in the

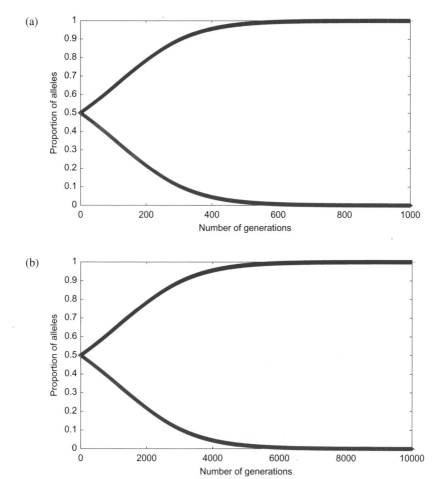

Figure 31.9   The proportion of alleles *A* (red line) and *a* (green line) with number
            of generations when the survival rates for *AA* and *Aa* (= *aA*) are equal
            to (a) 0.99 and (b) 0.999.

population we can calculate how the proportions of *A* and *a* vary in
successive generations. The result is shown in Figure 31.9(a). It will
be seen that after 1,000 generations, say 10,000 years for many animals, the population is virtually free from the characteristic corresponding to *A*. The modification of the genetic structure of the
organism is remarkably fast. If we change the 99% survival rate to

99.9% survival then the outcome is as shown in Figure 31.9(b). The almost complete removal of the allele A is slower in this case but still happens on a relatively short timescale.

The above example may apply to a situation where the conditions of life change — say, the average annual temperature falls — and advantage is obtained by those individuals in the population better able to survive that change. Another kind of evolutionary influence occurs because of *mutations*, or a change in the structure of DNA or another genetic material, RNA (ribonucleic acid). This can come about because of copying errors when DNA reproduces itself or because of damage, either by chemical agents or by radiation of one sort or another, say through excessive exposure to the Sun. Mutations cause variations in the gene pool of a species but, in general, mutations are harmful and are subsequently removed from the gene pool by natural selection. However, once in a while, and rarely, a mutation can be beneficial to the survival of the organism and in this case natural selection will eliminate the original gene and substitute the advantageous mutated gene.

We end our description of natural selection with the story of the peppered moth, *Biston betularia*. In the early part of the 19th Century this moth appeared in a range of shades of light grey with just the occasional darker individual. In 1848 a coal-black individual was observed near Manchester, presumably a mutant form, and by 1950 about 90% of the population was black. The reason for this process of natural selection was the industrial revolution, which produced vast amounts of pollution that blackened many surfaces, including the bark of trees. When trees had a pale brown or grey bark then the light grey peppered moths were less conspicuous on that bark and were less likely to be spotted and eaten by birds. However, when the bark of the trees turned black the pale moths were easily spotted by the birds whereas the mutant black forms were less conspicuous. Natural selection then ensured that the mutant form became dominant.

Chapter 32

Twentieth Century Genetics — The Alphabets of Life

32.1 Symbols and Alphabets

Expressing a message in terms of written language can be done in many ways. At one extreme there could be a separate symbol for every word; since there are about 200,000 words in the *Oxford English Dictionary* this would also be the number of symbols, although a few thousand of them would cover most everyday communications. The Chinese language comes closest to this with a full set of 80,000 characters, although 3,500 of them cover 99.5% of normal usage. Characters can be combined together to express subtle ideas so that, for example the characters for simple concepts such as 'much', 'clever' and 'man' could together convey the idea of 'genius'. However, there are other, and more efficient, ways of communicating in writing.

When we enter a library we see shelves stocked with a vast number of books, each with a different message to relate. Shakespeare's plays are richly varied and reading one of them tells you nothing about the next one you will read — after all, one play may be a tragedy and the next a comedy. Yet all the books in the library, including all of Shakespeare's plays, have something in common. They are all written using just the number of letters of an alphabet — 26 for English — with the subsidiary use of punctuation and the occasional inclusion of numerical digits and other symbols. We use the power of combining the letters in various ways to express words with different meanings and thus to create a virtually infinite range of messages with a finite, and small, number of basic components. The letters of the alphabet phonetically reproduce the spoken word — although many non-English speakers, learning the language and faced with the task of trying to write in English, might claim that the language has only a tenuous connection with phonetics!

In the early days of telegraphy the only signal that could be transmitted was a uniform frequency, and the only possible variation in such a signal was its timing. In the 1840s, an American, Samuel Morse (1791–1872), devised a scheme for telegraphic communication that depended on a sequence of either short or long signals, known as dots and dashes, respectively. Thus an 'a' is '• –', an 'e' is '•' and a 'p' is '• – – •'. A dash has the length of three dots, the space between the components of the same letter is one dot-length, the space between two letters is three dot-lengths and the space between two words is five dot-lengths. In this way just three components — a dot, a dash and a space — can be used to transmit the full range and subtlety of a language.

32.2 Proteins and the Protein Alphabet

Most people are aware of the term 'protein' as an essential component of a healthy diet. Meat and fish are rich in proteins and are a primary source for most people: vegetarians must take care to include high-protein vegetables in their diets — beans, spinach and broccoli for example. The reason for this requirement of a sufficient protein intake is that the body requires them for virtually every aspect of its structure and function. Proteins are required to build and maintain the muscles that enable a person to do physical work. Equally, they are necessary for the efficient working of the brain; an essential component of the activity of the brain involves the transmission of signals, enabled by the passage of small molecules, and also sodium and potassium ions, through membranes made of proteins. For healthy hair and nails a protein called α-keratin is required, which is also a component of feathers on birds, wool on sheep and horns, antlers and hooves on many herbivores. When the body is under attack by harmful bacteria and viruses, they are combated by antibodies, which are proteins. Oxygen is conveyed from the lungs to muscles and carbon dioxide from muscles to the lungs by the protein haemoglobin that resides in red blood cells. Other proteins are enzymes such as pepsin, which assists the digestion of food, the antibacterial agent lysozyme, which is present in tears, saliva, mucus and human milk, and also the hormone insulin, which regulates the level of blood sugar. This list of the role of proteins in humans, and in animalia in general, is by no means complete but is sufficient to underline their importance in the maintenance of life.

To understand how the body constructs proteins we must first know the general form of these important molecules. The basic units of which proteins are constructed are amino acids, 20 of which are involved in the structure of proteins — the protein alphabet. These are listed in Table 32.1. Chemists can deduce from the structural formulae, as given in the table, what the bonding arrangement is in each amino acid. Two of them contain sulphur (symbol S), otherwise only hydrogen (H), carbon (C), nitrogen (N) and oxygen (O) are involved; Ph represents a phenyl group, which is a hexagon of carbon atoms with hydrogen atoms attached.

A protein consists of individual amino acids linked end-to-end to form a long chain. To see how this is done we first show two amino acids, serine and aspartic acid, in Figure 32.1. All the amino acids contain the groups NH_2-C-COOH that are seen running along their centres. To link serine and aspartic acid as part of a protein chain the parts indicated in the dashed rectangles are lost (together equivalent to H_2O, i.e. water) and the linkages are established as in Figure 32.2.

A schematic unrealistically short protein is shown in Figure 32.3 with one unit of the protein chain indicated within the dashed box. The letter R represents a *residue* corresponding to one of the amino acids. At the two ends of the protein the arrangement of atoms is that appropriate to an isolated individual amino acid and they seal off the protein to give it stability. In the human body there are about one hundred thousand proteins, all performing different roles.

Table 32.1   The 20 amino acids in proteins, their structural formulae and abbreviations.

Amino acid	Formula	Abbreviation
Alanine	CH_3-CH(NH$_2$)-COOH	ala
Arginine	HN=C(NH$_2$)-NH-(CH$_2$)$_3$-CH(NH$_2$)-COOH	arg
Asparagine	H$_2$N-CO-CH$_2$-CH(NH$_2$)-COOH	asn
Aspartic acid	HOOC-CH$_2$-CH(NH$_2$)-COOH	asp
Cysteine	HS-CH$_2$-CH(NH2)-COOH	cys
Glutamic acid	HOOC-(CH$_2$)$_2$-CH(NH$_2$)-COOH	glu
Glutamine	H$_2$N-CO-(CH$_2$)$_2$-CH(NH$_2$)-COOH	gln
Glycine	NH$_2$-CH$_2$-COOH	gly
Histidine	NH-CH=N-CH=C-CH$_2$-CH(NH$_2$)-COOH	his
Isoleucine	CH$_3$-CH$_2$-CH(CH$_3$)-CH(NH$_2$)-COOH	ile
Leucine	(CH$_3$)$_2$-CH-CH$_2$-CH(NH$_2$)-COOH	leu
Lysine	H$_2$N-(CH$_2$)$_2$-CH(NH$_2$)-COOH	lys
Methonine	CH$_2$-S-(CH$_2$)$_2$-CH(NH$_2$)-COOH	met
Phenylalanine	Ph-CH$_2$-CH(NH$_2$)-COOH	phe
Proline	NH-(CH$_2$)$_3$-CH-COOH	pro
Serine	HO-CH$_2$-CH(NH$_2$)-COOH	ser
Threonine	CH$_3$-CH(OH)-CH(NH$_2$)-COOH	thr
Tryptophan	Ph-NH-CH=C-CH$_2$-CH(NH$_2$)-COOH	trp
Tyrosine	HO-Ph-CH$_2$-CH(NH$_2$)-COOH	tyr
Valine	(CH$_3$)$_2$-CH-CH(NH$_2$)-COOH	val

Proteins can vary in size from having a few tens of amino acids to having many thousands. For example, the protein haemoglobin, the component of blood that is responsible for transporting oxygen and carbon dioxide through the body, consists of an assemblage of four identical protein chains, each of which contains 287 amino acids.

With this background we can now give a general outline of the process that gives rise to the proteins that are needed both to create life and to sustain it.

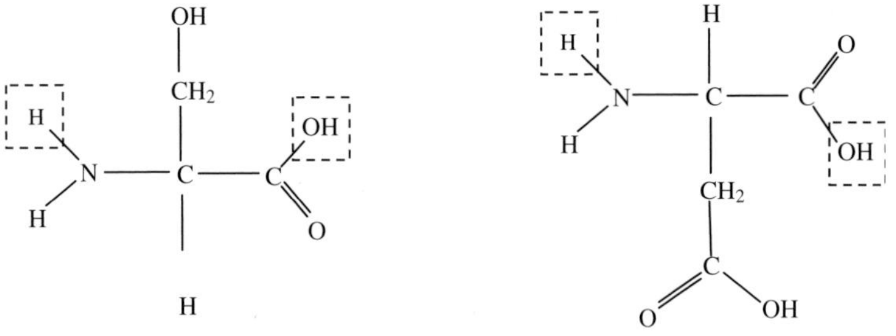

Serine

Aspartic acid

Figure 32.1 The amino acids serine and aspartic acid.

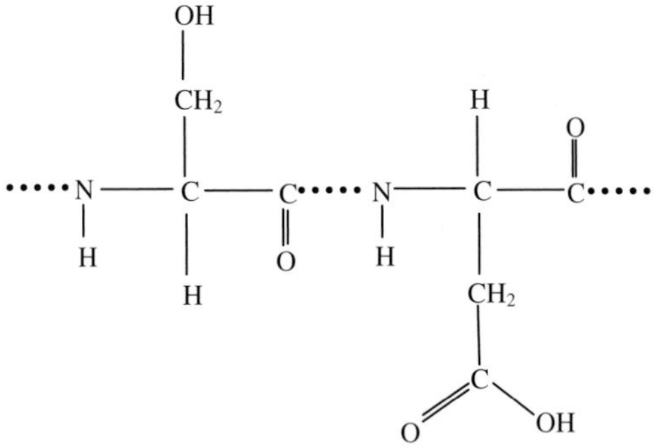

Figure 32.2 The linkage between serine and aspartic acid in a protein chain. The heavy dotted lines show the bonds between two amino acids.

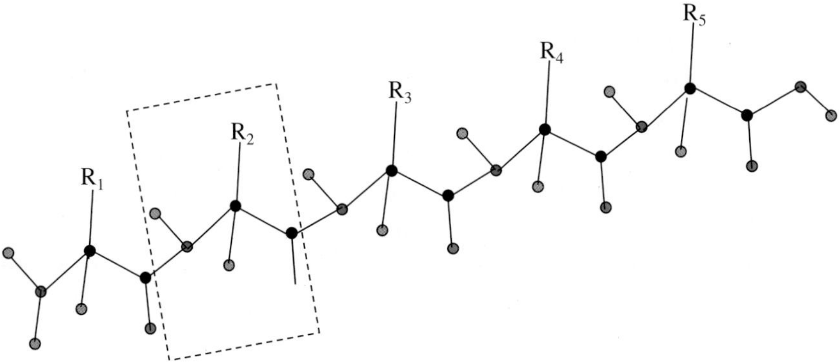

Figure 32.3 A hypothetical protein chain. Carbon ●, Nitrogen ●, Oxygen ●, Hydrogen ●.

32.3 The DNA Alphabet

Just as one can have a huge range of complexity of written material, from the simple message to the milkman, 'Two litres, please,' to novels of over one million words, so there is a huge range of complexity of living systems, from a single-cell bacterium to a human being. We have noted that all language communication, however simple or however complex, can be expressed in terms of a small number of symbols, primarily those of an alphabet. For proteins, it turns out that, similarly, all the vast number of proteins that exist can be defined in terms of 20 chemical units — a chemical alphabet in this case. The parallel between life and literature cannot be pushed too far; books do not have to reproduce themselves but living entities must do so. There are requirements of the chemical alphabet that an alphabet of literature does not have to meet.

The brief description of a bacterium in Section 30.1 gives no account of the detailed chemistry that goes on within the cell. However, despite the enormous complexity of the processes that occur to sustain a living organism one basic fact emerges — at the beginning of all these processes is DNA (deoxyribonucleic acid) that *completely* defines the organism and the way that it operates. To make an analogy with the world of engineering, it is the blueprint that defines the final form of the organism — a blueprint that needs some subsequent practical steps to create the object it defines. A blueprint in engineering, or in biology, can relate to something as simple as a stool or a bacterium, or something as complex as a Rolls-Royce aero-engine, or a human being. In the award-winning film *Jurassic Park*, scientists extract the DNA of long-extinct dinosaurs from mosquitoes trapped in amber, which had ingested the blood of those prehistoric creatures in the distant past. By introducing this DNA into the eggs of modern birds they bring the extinct creatures back into existence. The film is pure science fiction but the basis of it, that any living organism is completely defined by its DNA, is completely valid.

The surprising thing about DNA is that it is so simple, but then we have the analogy of the Morse code, dots, dashes and spaces with which we can express both a shopping list and the thoughts of the philosopher Ludwig Wittgenstein. Again, the simplicity of the components of a building gives no limit to the complexity of what can be built with it. St Paul's Cathedral, reduced to its elements, consists of blocks of stone, pieces of wood and a few other mundane and uninspiring materials, which could also be used to build a hut if one wished to do so. In the case of DNA there are just four basic units — called *nucleotides* and shown in Figure 32.4 — from which it is constructed. The common parts of each unit are ringed and are a phosphate group (a phosphorous atom surrounded by four oxygen atoms) and a sugar, *deoxyribose*. The component that is different from one nucleotide to another, referred to as the *base* of the unit, is one of the two *purines, guanine* (G) and *adenine* (A), or one of the two *pyrimidines, thymine* (T) and *cytosine* (C). These units can chemically link together in a long chain by the chemical bond, shown dashed, that links a sugar in one unit to a phosphate group in the next. The bases can appear in any order in the polymer — for example, ...AATGCGTAAG..... or ......AGACATAG...... — and this order defines both the nature and the detailed features of the organism. The sequence of letters, the instruction book or blueprint for the construction of the complete organism, is known as its *genome*. A particular sequence in part of the genome that gives the instructions for making a specific protein is known as a *gene* and this may involve anything from one thousand to one million bases.

The complete set of nucleotides for the whole organism is not linked together as one long chain but is arranged in several bundles of genes, each of which is a *chromosome*. The term chromosome is of Greek origin and means 'coloured object'. The chromosome is accompanied by proteins, the combination of chromosome and proteins being the chromatin observed by Walther Flemming (Section 31.2). Through chemical linkages with the chromosome, the proteins enable it to coil up in a stable configuration that can fit inside a cell; it is the whole chromatin that absorbs the dyes and so can be seen as highly chromatic objects when viewed in a microscope. Humans have 23 pairs of chromosomes, as shown in Figure 32.5. All but one of the pairs of chromosomes are similar in males and females, the exception being chromosome 23. For males, as shown in the

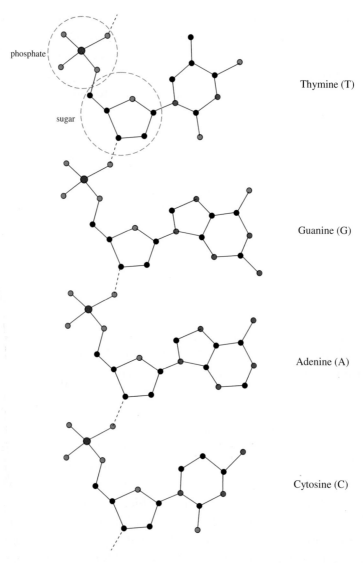

phosphate

sugar

Thymine (T)

Guanine (G)

Adenine (A)

Cytosine (C)

Figure 32.4   The nucleotides that form DNA, each consisting of a phosphate group, sugar and base. The linkage of nucleotides is shown by the dashed chemical bond. Hydrogen atoms are not shown. The other atoms are:

Carbon ●   oxygen ●   nitrogen ●   **phosphorus ●.**

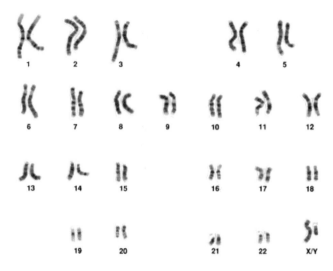

Figure 32.5 The 23 pairs of chromosomes for a male human.

figure, the members of the pair are different, one indicated as X and the other as Y. For a female both are X. It is this chromosome that controls the sex of a baby. The mother can only provide an X chromosome in her egg; if the father's sperm provides another X then the child will be female but if he provides a Y then the child will be male. When the English king Henry VIII divorced or executed wives for not bearing him a male heir, little did he know that his wives were blameless and that he was at fault!

The human genome contains about three billion bases organised into about 20,000 genes and 23 pairs of chromosomes. The largest human chromosome contains 220 million bases. One of the great triumphs of science is the mapping of the complete human genome in the early years of the 21st Century, achieved by the collaboration of scientists from many countries.

To give an idea of the relationship of genes to chromosomes we show a hypothetical pair of chromosomes in Figure 32.6 that contain ten genes. The alleles of the ten genes are labelled (A,a) to (J, j) and it will be seen that in some cases both chromosomes contain the same

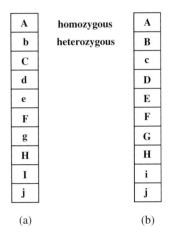

Figure 32.6   A pair of hypothetical chromosomes containing ten genes.

alleles, in which case for that gene the organism is homozygous, or they are different, giving heterozygosity for that gene. One or other of the genes in the pair of chromosomes will be passed on to progeny and each has a probability of 0.5 of being passed on. This is the basis for Mendel's interpretation of his pea experiment, although, of course, he was completely unaware of the nature of genes.

As a final comment on the nucleotide sequences in DNA, it is found that large parts of it have no known function in achieving the eventual goal of producing proteins — although, since nature seems to have a purpose in most entities it produces, there probably is some function. This 'non-coding DNA', sometimes called 'junk DNA', comprises about one half of the total DNA chain.

32.4 Determining the Structure of DNA

Before tracing the path by which the information in a DNA chain is converted into the proteins that give the organism its physical structure and function, we now examine the form in which DNA exists, because this is a vital factor in its ability faithfully to propagate the species. The complete information for the organism is coded within its DNA, which determines not only what species of organism is involved — i.e. an oak tree or a human being — but also the characteristics of that organism — i.e. brown eyes or blue. It turns out that DNA does not exist as a single polymer strand but as a double strand in the form of a helix. Indeed the term 'double helix' is well known to the public at large, even to those who do not understand its significance; its discovery was one of the great scientific events of the 20th Century. Four important characters played a role in this discovery — James Watson (b. 1928), Francis Crick (1916–2004), Maurice Wilkins (1916–2004) and Rosalind Franklin (1920–1958), shown in that order in Figure 32.7.

The story involved two laboratories. The first was the Cavendish Laboratory in Cambridge where Crick and Watson were engaged in building trial models of DNA using metal rods welded together to represent the bonds between atoms in parts of nucleotides, which could be flexibly joined together with barrel connectors to create a variety of configurations. The second was at King's College London, where Wilkins and Franklin were approaching the problem from a completely different direction. They had prepared samples of DNA in the form of fibres that had some of the characteristics of a crystal. When a beam of X-rays is directed at a crystal, and the crystal is spun around an axis, the incident X-ray beam is split up into a large number of beams coming out of the crystal in many directions with each beam having a different intensity. This process is known as X-ray diffraction and from the intensities and directions of the scattered beams it is

Figure 32.7   The cast in the 'Double Helix' drama (Maurice Wilkins photograph courtesy of Mrs Patricia Wilkins).

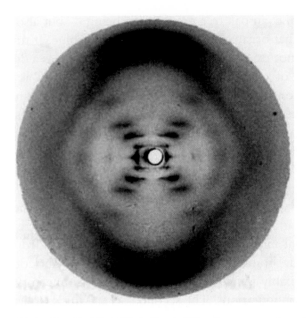

Figure 32.8   Rosalind Franklin's X-ray diffraction pattern from DNA.

possible, in principle, to find the arrangement of atoms in the crystal — although, in practice, it is usually extremely difficult to do so. Because the DNA fibre was only partially crystalline, instead of sharp diffracted beams, which could be recorded on film as small spots, the diffraction pattern was rather smeared (Figure 32.8). This picture may seem to be devoid of interest or information but it was a tremendous achievement by Rosalind Franklin since nobody had produced anything as good previously.

On a visit to King's College, Crick and Watson were shown Rosalind Franklin's X-ray picture by Maurice Wilkins. To the Cambridge pair, with their background of building a whole succession of models, the information that the picture gave was the key to the problem of solving the DNA structure. The X at the centre of the picture told them that DNA was in the form of a helix and the horizontal streakiness told them not only that the bases in DNA were all parallel to each other but also how far apart they were. There were other clues, apart from those in the photograph, which helped

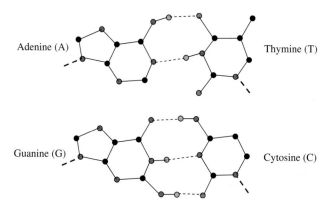

Figure 32.9   The linking of the base pairs A + T and G + C by hydrogen bonds, shown as faint dashed lines. The heavier dashed lines are chemical bonds to the phosphate-sugar chain.

Crick and Watson to produce their model. It had been known for some time that when different samples of DNA were chemically analysed the amounts of the bases adenine and thymine (A and T) were the same, as were the amounts of guanine and cytosine (G and C). In each case one of the pair was a purine and the other a pyrimidine. Watson, using cardboard models, showed that pairs A + T and G + C, the chemical structures of which were planar, could chemically bind together with linkages known as hydrogen bonds in which hydrogen atoms act as a kind of glue holding the bases together. The bonding of A + T and G + C is shown in Figure 32.9. Clearly the reason for the equality in amounts of A and T and of G and C was because they were always linked together in pairs.

All the information was now to hand for Crick and Watson to build a model that quickly convinced the scientific community that the structure of DNA had indeed been solved. One begins with two parallel linked backbones of phosphate + sugar, bound in long chains by bonds as shown in Figure 32.3. Now the bases are attached in an order that corresponds to the characteristics of a particular organism. Where there is A in one chain then opposite, in the other chain, is T and these are bound together by hydrogen bonds. Similarly C always appears opposite to G. Next the chains are twisted into a helical form, which explains the X in Rosalind Franklin's X-ray photograph. This twisting is done in such a way that the planes of the bases are all parallel to each other and perpendicular to the axis of the helix — which explains the streakiness in Figure 32.8. It sounds very simple, but Crick and Watson had to use a great deal of ingenuity and their experience in model building to achieve the final result.

The structure was published in the journal *Nature* in 1953 and Crick, Watson and Wilkins were awarded the Nobel Prize for Medicine in 1962. Sadly, in 1958 Rosalind Franklin died at the early age of 37 so her important contribution could not be considered in the allocation of the prize. However, her contribution is widely recognised and every year the Royal Society makes the Rosalind Franklin Award to a female scientist or engineer for an outstanding contribution in some area of science, engineering or technology.

An impression of part of the structure of DNA is given in Figure 32.10. The helical structure corresponding to the phosphate + sugar backbone is clearly seen as is the system of linked base pairs that, in their entirety, describe the organism to which they belong.

What is so significant about the double-helical structure of DNA? After all, it is just the message contained in one strand that contains the essential information about the organism, so the other strand seems redundant. Well, in fact, the helical structure is a *critical* factor in the whole mechanism by which a species continues its existence. If an individual organism is to grow, or faithfully pass on its genetic structure to a new generation, then it is necessary to reproduce new DNA from existing DNA *without producing any errors in the reproduction process*. This happens in the following way. A helical section of DNA unravels into its two component strands in the presence of a source of the different nucleotides. Each base of the unwound part of the strand then attaches to itself a nucleotide corresponding to its base-pair partner. As the DNA helix unwinds so each of the original strands build up a partner strand, preserving the base-pair relationship, and rewinds itself into a helix. After the original helix has

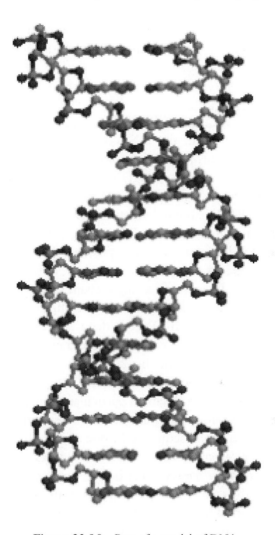

Figure 32.10 Part of a model of DNA.

completely unwound, the end product is two DNA helices in place of the original one and each is a *precise* copy of the original. In essence, each strand of the DNA acts as a template for the formation of a partner strand and, in principle, this copying process can be repeated indefinitely.

The next step in the story of DNA is to describe how it leads to the proteins that, in their form and function, define different organisms.

32.5 The Role of RNA

The message contained in the DNA base sequence has to be converted into proteins but, just as a set of instructions written on paper cannot by itself make a stool, so DNA on its own cannot make a protein. The process of going from DNA to proteins requires an intermediary — RNA (ribonucleic acid), which is a long chain compound with some similarities to DNA. The differences are that it contains the pyrimidine uracil (U) in place of thymine, contains the sugar ribose (the deoxyribose in DNA is ribose with an oxygen atom missing) and normally exists as a single chain and not in the double helix form of DNA; RNA does not have to replicate itself so a double helix form is unnecessary. If DNA is described as a blueprint then RNA is correspondingly a computer program that controls the machine tool that makes the final product — the proteins that together create the final organism.

In the first stage of producing a particular protein, RNA must be produced from the information in a gene corresponding to that protein, expressed as a DNA sequence in one of the chromosomes. An enzyme, *RNA polymerase*, attaches itself to a region of the DNA known as the *promoter sequence* just before the gene of interest. The DNA sequence then unwinds and an RNA strand is made as a complementary version of the nucleotide sequence in the gene, with uracil in RNA as the complementary partner to adenine in DNA and other corresponding pairs as adenine partnering thymine, cytosine partnering guanine and guanine partnering cytosine. The DNA also includes a sequence of nucleotides that indicates when the construction of the RNA strand should be terminated.

This process is known as RNA *transcription* but, since there is a one-to-one correspondence of what is in the RNA and the DNA sequence, then non-coding parts of the DNA are transcribed into non-coding parts of RNA. The next stage in the process is to strip out the non-coding parts of the RNA, which is performed by other specialist RNA molecules, and the end product is *messenger RNA*, mRNA, which is the complete instruction set for producing a protein without the encumbrance of any unnecessary components. The mRNA contains the required information in a very straightforward way. Sequences of three nucleotides, from A, U, G and C, known as *codons*, indicate the next amino acid to add to the protein chain and also when to terminate the chain. The correspondence between codons and amino acids is given in Table 32.2.

Once the mRNA has been produced it leaves the nucleus of the cell and attaches itself to the ribosomes, seen in Figure 30.3, where it controls the construction of the protein appropriate to the gene that initiated the process. The way that this is done is represented schematically in Figure 32.11. An important element of this process is the presence of *transfer RNA* (tRNA) molecules, which are short sequences of about 80 nucleotides. An amino acid is attached to a tRNA molecule, which has three of its consecutive nucleotides

Table 32.2   The amino acid (or termination) equivalents of the 64 codons.

Amino acid	Codons	Amino acid	Codons
Alanine	GCU GCC GCA GCG	Lysine	AAA AAG
Arginine	CGU CGC CGA CGG AGA AGG	Methonine	AUG
Asparagine	AAU AAC	Phenylalanine	UUU UUC
Aspartic acid	GAU GAC	Proline	CCU CCC CCA CCG
Cysteine	UGU UGC	Serine	UCU UCC UCA UCG AGU AGC
Glutamic acid	GAA GAG	Threonine	ACU ACC ACA ACG
Glutamine	CAA CAG	Tryptophan	UGG
Glycine	GGU GGC GGA GGG	Tyrosine	UAU UAC
Histidine	CAU CAC	Valine	GUU GUC GUA GUG
Isoleucine	AUU AUC AUA	Terminate	UAA UAG UGA
Leucine	UUA UUG CUU CUC CUA CUG		

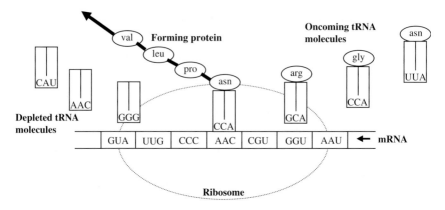

Figure 32.11   The building of a protein using an mRNA template and amino acid
components delivered by tRNA.

giving a complementary codon to that for the attached amino acid;
the other nucleotides in tRNA are just carriers for the important
three that form the codon. Thus, for example, one of the codons in
mRNA for arginine is CGU so the complementary codon for tRNA
is GCA. In Figure 32.11 the ribosome is seen 'reading' a codon
from the mRNA with sequence AAC, corresponding to the amino
acid asparagine. It has bound to a tRNA molecule through the
complementary bases CCA and the tRNA will release asparagine to
join the growing protein chain. When the complete protein has
formed, the mRNA will deliver one of the terminating codons,
UAA, UAG or UGA, and a tRNA with a complementary codon,
AUU, AUC or ACU, will bind itself to the mRNA and will terminate the chain in the way indicated in Figure 32.3. Of course, this
is a rather general description of the protein-forming process and a
great deal of very complex chemistry is happening to make the process work, but the essential steps in the process are as described
here.

32.6 Sexual Reproduction

In the process of developing an infant from a foetus or in growing
from infancy to maturity it is necessary for cells to multiply to

provide the substance of the growing organism and, since cells have a limited lifetime, even in the mature adult, replacement of cells is necessary. The process of cell division, in which the new cells are exact replicas of the originals, is called *mitosis* and is illustrated in Figure 30.4. However, in sexual reproduction a combination of parts of the genetic content of the cells of the two parents produces an individual with a mixture of the individual characteristics of the parents, a process known as *meiosis*. This method of reproduction, which ensures that the offspring are different from both parents, combined with the process of natural selection, ensures that the species endures robustly and does not degenerate. It also allows for slow variation that gives adaptation to changing circumstances and that, over the longer term, can lead to a new species. A representation of the process of meiosis is given in Figure 32.12, where the description is restricted to one type of chromosome, say number 4 in Figure 32.5. The same happens for all 23 chromosomes that

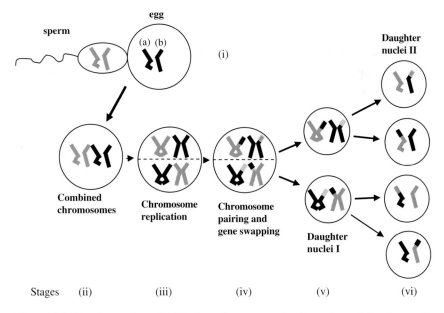

Figure 32.12   Stages from fertilisation of an egg to the formation of daughter cells with a different mix of genes.

together give the human genome. There are several stages in the process.

(i) The sperm penetrates the egg.
(ii) The egg now contains two pairs of chromosomes, one pair, (a) and (b) (see Figure 32.6), from the male (lighter grey) and the other pair from the female.
(iii) DNA replication produces a second copy of each of the four individual chromosomes. Like pairs, e.g. the male (a)s, link together and the male (a) pair and female (b) pair become separated from the male (b) and female (a) pairs. The separation is shown as a dashed line.
(iv) In each separated group, genes are swapped in a random fashion between chromosomes in the (a) pairs and (b) pairs, e.g. a gene in one of the male (a) chromosomes will swap with the corresponding gene in one of the female (b) chromosomes.
(v) The egg divides to form two 'Daughter I' cells, each containing one of the separated male-plus-female pairs of chromosomes.
(vi) The linked pairs of similar chromosomes break apart and each of the two cells divides to produce two cells (now four Daughter II cells in all) such that each Daughter II cell contains an (a) and a (b) chromosome and also a contribution from the male and the female.

This seems to be a complicated process, and needs study of the figure to understand thoroughly, but what it ensures is that each of the four Daughter II cells contains a random mix of contributions from the two parents — the essential condition for Darwinian selection to occur. Any one of the four Daughter II cells that survives can go on to further multiply by mitosis to form first a foetus and finally a fully-developed infant.

The description we have given here of the pathway from DNA to the formation of a new living organism is very broad-brush and leaves unexplained many of the marvels of how the processes occur. The human embryo is a collection of a few cells containing a random mixture of the parental genes. The cells multiply and start to differentiate so that they produce the different organs of the infant that they eventually become. Some cells become left-arm cells and some become right-arm cells and, unless some terrible error occurs, the infant will have two arms and not one or three. For this process to happen properly, the chemistry of the mother's body must be just right. In the 1960s a new sedative drug, thalidomide, was introduced into the British pharmacopoeia and was prescribed to some expectant mothers to treat stress and morning sickness. Subsequently large numbers of babies were born with stunted limbs; clearly the thalidomide had interfered with the system that controlled the formation of the infant in the womb. It was believed that the placenta, which prevents the mother's blood from mixing with that of the foetus, would also be a barrier to prevent drugs taken by the mother from affecting the unborn child — but clearly this was not true.

We may know something about the relationship between DNA, chromosomes and genes and the final organism that is produced, but there is still much to learn and understand about the mechanisms that operate to produce that relationship.

Chapter 33

Life Begins on Earth

33.1 Early Ideas on the Origin of Life

And God said, Let us make man in our image, after our likeness: and let them
have dominion over the fish of the sea, and over the fowl of the air, and over the
cattle, and over all the earth and over every creeping thing that creepeth upon
the earth.

And the Lord God formed man of the dust of the ground, and breathed into
his nostrils the breath of life; and man became a living soul.

And out of the ground made the Lord God to grow every tree that is
pleasant to the sight and good for food.

And the Lord God caused a deep sleep to fall upon Adam, and he slept;
and he took one of his ribs, and closed up the flesh instead thereof; and the rib,
which the Lord God had taken from man, made he a woman.

The lines above are extracts from the biblical book of Genesis that
describe the creation of living organisms, an event that can be dated
by biblical analysis to less than 6,000 years ago. The scientific evidence is against both the time and form of this process of creating life,
and modern religious interpretation is that this description is a meta-
phor for the process of creation that, whatever its exact nature, was of
divine origin. However, there are fundamentalists within the faiths of
Judaism, Christianity and Islam who believe that the Genesis description is literally true.

The observation that very often life appeared apparently as the
result of decay, for example, maggots being produced by rotting
meat, led to the early belief that life rose spontaneously from rotting

materials — and, despite evidence to the contrary, some people believed this right up to the beginning of the 19th Century. The evidence against spontaneous biogenesis was finally produced in 1668 by an Italian doctor and naturalist, Francesco Redi (1626–1697). He carried out experiments in which portions of fresh meat and fish were put in jars, some left open to the air and others covered with gauze, so that air could enter the jars but nothing else. After a few days maggots appeared in the uncovered jars but not in those covered by gauze, showing clearly that the maggots had not arisen spontaneously but were the result of flies laying eggs on their contents. Later experiments, carried out by Louis Pasteur (1822–1895) in 1861, extended the evidence against spontaneous biogenesis to the microscopic world by showing that bacteria and fungi do not appear in sterile conditions, even in nutrient materials favourable to their growth.

Since life had been shown not to be generated spontaneously, the problem then became how it originated from non-living matter — a problem that has still not been convincingly solved.

33.2 The Origin of the Components of the Molecules of Life

When we consider the complexity of life, involving not only the 'blueprint' materials of DNA but also various forms of RNA (there are more than the three that played a part in the simplified narrative in Section 32.5) and the complex mechanisms required to translate those blueprints into living material, then we may well despair of ever finding a scientific explanation for how even the simplest organism could form. The problem of forming life can be broken down into a number of stages. The first is that of producing the necessary chemicals although what 'necessary' means in this context is open to question. Some nucleotide chain, either DNA or RNA, would seem to be indicated, together with some basic proteins that would be necessary to activate the chain in some way. The next stage would be to assemble the chemicals into a self-replicating, but at that time not necessarily living, system because the most essential requirement of life is its own maintenance and continuity. The final step would be to construct a single-cell entity with the basic DNA–RNA–mRNA–tRNA–protein mechanism for growth, maintenance, propagation and potential evolution. Each of these stages presents formidable, seemingly insurmountable, challenges for achievement by spontaneous, and presumably random, events — but life is a fact so these things must have happened somewhere, somehow.

Along with the formation of the Universe, it has the status of the most challenging problem faced by scientists today. It has preoccupied many leading scientists, nearly all of whom have limited themselves to considering environments within which the essential-for-life chemical compounds could form. In 1924 an eminent Russian biochemist, Alexander Oparin (1894–1980), suggested that, in the early Earth with an absence of oxygen, materials such as methane (CH_4), ammonia (NH_3), water (H_2O) and hydrogen (H) would have come together to form complex molecules that would be able to replicate themselves, albeit they would not be 'living' in the full meaning of the word. The most famous early experiment to test Oparin's ideas about how basic life chemicals could have originated was carried out in 1953 by two Americans, Harold Urey (1893–1981) and Stanley Miller (1930–2007). They put water into a container to which they added the gasses that were proposed by Oparin and would be expected in the primitive atmosphere of the early Earth — methane, ammonia and hydrogen. Then electrical discharges were passed through the gasses to simulate the passage of lightning. When Urey and Miller examined the final contents of the container they found many organic compounds, including amino acids, the components of proteins. There is evidence from the detection of compounds within the atmospheres of the major planets and of the gasses coming from comets that other carbon-containing compounds could also have been available in the early Earth. An experiment carried out in 1961 by the Spanish-American chemist Juan Oró (1923–2004) showed that amino acids could also be made from hydrogen cyanide (HCN) plus ammonia in an aqueous solution. Of even greater significance, his experiment yielded a significant quantity of adenine, one of the four bases in DNA, and later experiments showed that, under slightly modified conditions, the other three bases in DNA could also be produced. Another certain source of organic materials on Earth can be found in meteorites. One carbonaceous chondrite (Section 17.3) called Murchison, weighing over 100 kilograms, which landed in Australia in 1969, has been found to contain many organic materials, including 19 of the 20 amino acids that go into the formation of proteins.

33.3 The Origin of the Molecules of Life

There seems to be abundant evidence that the basic building blocks of life can arise by chemical reactions, starting with very simple starting molecules in the conditions of an early Earth, but they are just the components of the molecules of life — basic amino acids or the bases of nucleotides — rather than the molecules themselves. We now have to consider the conditions under which these components could assemble themselves into a chain[a] and several ideas have been advanced in this direction.

A novel and interesting idea was put forward in 1988 by Günter Wächtershäuser (b. 1938), a German patent lawyer, a chemist by education who developed an interest in genetic engineering. He suggested that good environments for the components of life chemicals to assemble themselves into chains were in underwater volcanic vents where there were hydrothermal flows at high pressure and temperature in the presence of many materials, both organic and inorganic. Under these conditions, in the presence of several metals that would act as catalysts,[b] reactions between potential early atmospheric gasses and iron sulphide (a fairly common material on Earth) could both produce more complex organic compounds and also release enough energy to promote the formation of short strings of amino acids or nucleotides. A difficulty with Wächtershäuser's idea is that the high-temperature conditions conducive to forming the basic organic

---

[a] We use the word 'chain' here rather than 'polymer'. Technically a polymer is of infinite length, although in practice the term is applied to a very long terminated chain.

[b] Catalysts are materials that promote chemical activity without being themselves consumed by those reactions.

compounds were also conducive to breaking up chains, so that any chains formed would have a very short lifetime. More recently, in 2002, two scientists, William Martin (b. 1957), an American scientist working in Germany, and the British scientist Michael Russell (b.1939), have proposed a more congenial environment for the Wächtershäuser process — in so-called *black smokers*, streams of gasses and chemicals from the Earth's interior emitted from vents on the sea floor. These streams contain hydrogen, cyanides, carbon dioxide and many other materials and they move amongst minute cavernous structures on the sea floor that are internally coated with iron sulphide. If any organic molecules are formed within these cavities they tend to be retained for some time, rather than immediately diffusing into the ocean, and hence have a greater chance of linking with other molecules. Another feature is that within the black smokers there are high temperature gradients; the higher-temperature regions would favour molecule formation while lower-temperature regions would both allow chain formation and give chain stability.

These are just some of a number of different ideas that have been advanced for explaining both the formation of the components of DNA, various forms of RNA and proteins, and how short chains of these components could have formed. They serve to illustrate how far we are from any convincing theory of the origin of life. If we take the optimistic view that a long DNA molecule could be produced by some process or other, then we would have to deduce how it became encased within a suitable material to constitute a cell and how the other essential components of a single-celled life form — RNA, mRNA, tRNA and ribosomes — could come about. Clearly there must be many steps between producing the chemicals and producing a living cell but, thus far, no description of this transition has been given.

33.4 The Source of Life

Although we have no idea at present how life began, we can, nevertheless, consider some general questions about the origin of life. The first of these is whether or not life would almost inevitably and spontaneously arise wherever the conditions — i.e. temperature and chemical environment — were suitable. The antithesis of this proposition is that life is so unlikely to arise that it would have occurred very few times within the Universe over its whole lifetime. We know that life occurred at least once since we, who are alive, are now considering that very question! That is a self-evident statement of the *anthropic principle* that states that any theories concerning the development of the Universe (including life within it) are constrained by the need for life to occur. The SETI (Search for Extra-Terrestrial Intelligence) project is an attempt by groups of scientists to detect radio signals that might have emanated from an intelligent source outside the Solar System. If such a signal were to be received then this might indicate that life was probably very common in the Universe; it would be extremely unlikely that two life forms would make contact if life was a rare phenomenon. Actually it is possible that life does occur frequently — perhaps arise inevitably under the right conditions — but either rarely reach a stage of development at which it could communicate or, having reached such a stage, would be short-lived. There are those who argue that it is only a hundred years or so since mankind established radio communication and that a catastrophe, perhaps due to nuclear weapons or global warming, is likely to terminate human life within the next few hundred years — a short period of time in cosmic and evolutionary terms. That may be a pessimistic view, but it is possible. It may seem paradoxical, but perhaps the evolution of high intelligence and the ability to develop advanced technologies may not be an asset in the struggle for survival!

Another lively topic is the search for life, either extant or extinct, on Mars, a planet where the temperature is within the range where life could exist and which is thought to have had a substantial aqueous atmosphere and a warmer climate in the past. The argument, previously given, that detecting life elsewhere would indicate that life was commonplace in the Universe, which is perhaps the basis of the SETI and Mars searches, is only valid if the life sources are of completely independent origin. If life had originated on Mars when conditions there were suitable, and was later transported to Earth in a simple unicellular form on meteorites, then the deduction that the formation of life is a relatively common phenomenon would collapse.

So far no signs of extra-terrestrial life have been detected. There is, however, an intriguing situation on Earth itself. In Section 30.3 we described archaea, organisms that resemble bacteria but which are not bacteria. They are present in environments of extreme temperature, salinity and alkalinity and are also present in deep underground locations. They may have originated on Earth when conditions were too hostile for any other life form to develop. When they were first discovered they were considered to be bacteria that had adapted to extreme conditions, but in the 1970s it was found from their DNA that they are a completely different life form from bacteria. This then raises the question of whence bacteria and archaea derived. As simple as they are they are still too complex to have been produced by just putting together the necessary chemicals to become the first forms of life. They must have had precursors, although not necessarily living precursors by the definitions we use for living organisms. Since they are so simple they must be close, in developmental terms, to the origin of life. Could they be derived from a common ancestor and, if so, how did their DNA compositions subsequently diverge so widely? If they did not derive from a common ancestor then this suggests that *there may have been two distinct life forms that evolved independently on Earth.* On this interpretation archaea developed and flourished under the harsh conditions of early Earth and, when conditions became more benign, as we understand that term, instead of archaea adapting to the new conditions a completely new life form arose, leading to bacteria. Such a situation, if it were true, would be as significant for assessing the probability that life arises spontaneously in suitable environments as would be the discovery of life on Mars. However, the prevailing view is that bacteria and archaea probably do have common ancestry.

There are those who believe that life did not begin on Earth at all but arose elsewhere and was then transported to Earth. There are two variants of this theory — *panspermia*, which proposes that the seeds of life exist throughout the Universe and may take root in many locations, and *exogenesis*, the idea previously mentioned, which proposes that life began elsewhere in the Solar System, say on Mars, and then was carried to Earth, presumably by a Mars meteorite. The panspermia idea was championed by the two British astronomers Fred Hoyle

(1915–2001) and Chandra Wickramasinghe (b. 1939) who proposed that the seeds of life were transported to Earth in comets and in dust that originated in comets and that, indeed, this process continues to this day. When the idea was first proposed, in 1983, it was largely discounted by astronomers, biologists and astrobiologists (scientists who are concerned with biological aspects of astronomy). However, many have had second thoughts and the idea is now at least respectable enough to be seriously considered. In 2001 an Indian experiment, involving a high-altitude balloon, collected dust from above the stratosphere at a height of 41 kilometres. The collection process was carried out in carefully controlled aseptic conditions to ensure that there was no contamination from terrestrial sources. When the dust was examined it was found to contain bacterial material (Figure 33.1). There is a small possibility that this was of terrestrial origin — perhaps ejected into the upper atmosphere by a volcanic eruption — but, on the whole, the evidence seems against that idea.

The panspermia hypothesis does not answer the question of how life began but it does remove the need for it to have occurred under terrestrial conditions and it does open up the whole Universe, or at least the Milky Way, as a possible source of life. Life is almost certainly

Figure 33.1   Spherical bacteria collected from a height of 41 kilometres (Indian Space Research Organisation).

the result of an extremely unlikely set of events and the probability that they happened in any particular site where the conditions were suitable would, in most circumstances, be regarded as zero. However, there are one hundred thousand million stars in our galaxy and there could be a similar, or even greater, number of possible sites for the formation of life including planets or bodies like comets. Even if the probability at any one site that life would develop there is one in a trillion (million million) then it is not unlikely that, somewhere, life would begin. As we have seen in Section 31.4, once life begins then there are pressures and processes that increase the complexity of life and, given enough time, may create intelligent forms of life, of which we humans may be a fairly primitive example.

The question of how life began is still unresolved, and may well be insoluble for human intelligence. Of one thing we can be sure and that is that life *did* evolve and that bacteria were around just one billion years after the Earth formed. Figure 33.2 shows a fossil of *cyanobacteria*, otherwise known as *blue-green algae*, dating back to that time. In the intervening 3.5 billion years, more complex life forms, including *homo sapiens*, came into being. We shall see in Chapter 35 that cyanobacteria played an important role in creating the conditions for most advanced life forms to exist.

33.5 The Creation of Self-Replicating DNA

In May 2010 a remarkable experiment was carried out that was on the threshold of creating a living entity. Scientists at the J. Craig Venter

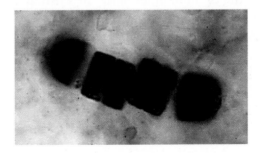

Figure 33.2   A fossil of blue-green algae, 3.5 billion years old.

Institute, led by John Craig Venter (b. 1946), an American biologist and business-man, created synthetic DNA and transplanted it into a bacterium, which then replicated itself, so reproducing the synthetic DNA. They first determined the genome of the bacterium *Mycoplasma mycoides*, next created a synthetic copy of the genome, for which standard techniques and synthesising equipment are available, and finally transplanted the copy into another bacterium, *Mycoplasma capricolum*. Once inserted into the new bacterium, the synthetic DNA modified or eliminated 14 existing genes in the host to suit its own development. To be sure that they would know that what was being replicated was the synthetic DNA, the scientists incorporated some non-functional nucleotides in their copy genome, and these were present in the replicating bacteria.

This work has implications of many kinds. It could open the door to many novel industrial applications — for example, designing algae that would intake carbon dioxide from the atmosphere and emit hydrocarbons that could be used as a biofuel, which would be a perfect green technology. Alternatively it could be used to design algae that could efficiently use light energy to break up water into its constituent hydrogen and oxygen — again a source of green fuel. However, such a development also raises ethical issues. Although in the Craig Venter experiment the synthetic DNA was a copy of what existed naturally, the question arises of what the consequences would be of producing an alien DNA, one that does not already exist in a terrestrial organism. If this could reproduce when planted in a host bacterial cell then a new life form would have been created — although it would not be the creation of life since the existing living cell and all it contained, except its original DNA, would be used. Could this create a dangerous bacterium to which animalia had developed no resistance and against which no antibacterial agents were effective? There are many who are concerned about the creation of synthetic life, the implications of which might now just be in the realm of science fiction, but which could conceivably become science fact.

Chapter 34

The Restless Earth

34.1 The Jigsaw-Puzzle Earth

In Section 28.2 we described the Earth at a stage where a solid surface had formed and the atmosphere was very different from that at present. Most of the gasses then in the atmosphere are now present only as very minor components, with the single exception of nitrogen, which forms 78% of the present atmosphere. The notable absence in that ancient atmosphere was oxygen, without which most life on Earth would be impossible; how oxygen became a substantial part of the present atmosphere is an important topic of Chapter 35.

We left our description of the Earth about 3,500 million years before the present time and if we could see that Earth we would not recognise it. It would be hot and steamy, containing both sea and land masses, but with the land not organised into the continents that we know today. In some of the least hostile environments some simple single-celled life forms had arisen — archaea and, perhaps, bacteria. We should certainly wonder how *that* world could have become the world we now inhabit.

In 1564 a Dutch cartographer, Abraham Ortels (Figure 34.1(a)), produced the first reasonably accurate map of the world as it was then known. In 1596, looking at the outlines of the continents in his map as shown in Figure 34.1(b), he suggested that America, Europe and Africa had once been joined together and had been pushed apart by 'earthquakes and floods'. We can see in his map, what he saw, and what others have subsequently seen, that the east coast of the

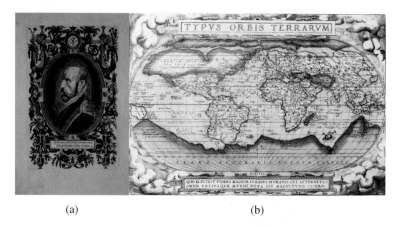

(a)   (b)

Figure 34.1   (a) Abraham Ortels (1527–1598) and (b) Ortels' world map, 1570.

Americas and the west coast of Africa appear to fit together like two pieces of a jigsaw puzzle. The idea that continents could move with respect to one another remained as a curious suggestion for 350 years, and then exploded on the scientific scene as a hotly disputed topic involving many eminent geologists.

34.2 The Evidence for Continental Drift

The idea that continents were once joined together and could somehow move apart was considered from time to time, but seemed so outlandish that there was reluctance by most scientists to espouse the idea in a formal way, for fear of being ridiculed. However, in 1912 a bold German geologist, Alfred Wegener (Figure 34.2), published a paper in which he formally proposed the idea that continents moved apart, using the term 'die Verschiebung der Kontinente' which translates as *continental drift*. According to Wegener, the process by which the continents move was that the land masses representing the continents moved through the ocean floor in the way that the blade of a plough moves through the surface of a field. The whole idea seemed absurd and the theory was derided by most of the leading scientists and geologists of the time. After all, how could chunks of a solid crust move around and what forces were available to make them move?

Figure 34.2   Alfred Wegener (1880–1930).

The feeling that continental drift could not happen was so strong that very eminent geologists, notably the leading British geologist Harold Jeffreys (1891–1989), *never* accepted the idea even when, later, the evidence for it became irrefutable.

Evidence soon began to accumulate that, however outlandish Wegener's theory seemed to be, it was actually true. For example, the fossil records of different continents showed the same plants and animals of the same age in what are now well separated locations. In Figure 34.3 there is shown the probable arrangement of land 250 million years ago when the southern continents formed a single super-continent, called *Gondwana*. The coloured bands in the figure indicate locations where common flora and fauna fossils have been discovered. Of equal importance is the similarity of rock types in the present continents corresponding to contiguous regions of Gondwana.

The evidence is overwhelming that the continents are now in very different positions from where they were hundreds of millions of years ago. Apart from the fossil record illustrated in Figure 34.3 there is the occurrence in Antarctica of coal, the fossil remains of tropical plants. Another indication of continental drift is striations in the surfaces of the

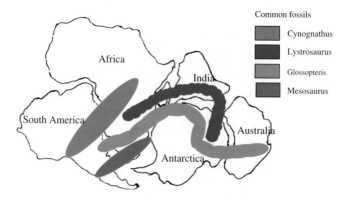

Figure 34.3   Regions of fossils of similar species in Gondwana. Cynognathus and Lystrosaurus are land reptiles from about 250 million years ago. Glossopteris is a type of fern and mesosaurus a fresh-water reptile.

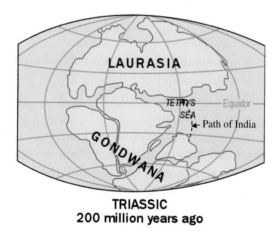

TRIASSIC
200 million years ago

Figure 34.4   The land mass of the Earth 200 million years ago (USGS).

southern continents, due to the motions of giant glaciers, which line up when the continents are assembled into Gondwana. The motion of these glaciers, with the continents in their present positions, tend to run from the equator southward, indicating clearly that the continents were once placed very differently with respect to the Earth's spin axis.

Putting all the scientific evidence together gives a picture of the land masses arranged somewhat as in Figure 34.4 about 200 million years ago. There is not complete agreement about the exact arrangement of land masses in the distant past and, indeed, given the fact of continental drift, the arrangement is time-dependent. Some representations of the original combined land mass give a lesser separation of the Americas by moving North America further south and this then makes the two continents, *Laurasia* and Gondwana, labelled in the figure, seem less distinctive. The combined land mass, Laurasia plus Gondwana, is usually referred to as *Pangaea,* a word derived from Greek meaning 'all the Earth'.

34.3 The Mechanism of Continental Drift

By the late 1950s the idea of continental drift was generally accepted by the scientific community, although there were a few diehards who never accepted the idea. What was needed was some theory to explain *how* continental drift happened, since Wegener's 'ploughing' model was obviously impossible. In fact, we have already described the mechanism that would provide the solution to this problem when we explained the form of the hemispherical asymmetry of Mars due to polar wander in Section 25.1. Although a solid layer had formed at the surface of the early Earth, it was not a *static* layer. This layer, known as the *lithosphere*, consists of the low-density rocks that constitute the crust of the Earth plus a solid layer of the mantle, the denser rocks below the crust. Below the lithosphere the mantle, although virtually solid, can flow very slowly like a liquid; this region is called the *asthenosphere*. We saw in Figure 28.2 that convection currents in the asthenosphere, which bring up heat from below and enhance the Earth's rate of cooling, also apply drag forces on the lithosphere that in some places tends to tear it apart and in other places tends to crush it together. Now we consider what happens in regions where the lithosphere is being pulled apart, shown in schematic form in Figure 34.5. The lithosphere moves outwards and magma wells up from the Earth's mantle to fill the gap. Eventually this solidifies to form a new section of lithosphere that continues to move outwards. A ridge forms on either side of the rift.

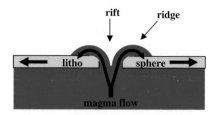

Figure 34.5 Creation of new lithospere in the gap when the exiting lithosphere is torn apart.

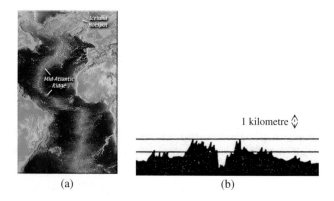

(a)   (b)

Figure 34.6 (a) The mid-Atlantic ridge (USGS) and (b) a cross section of the ridge (vertical and horizontal scales differ).

An example of a structure formed on the Earth in this way is the mid-Atlantic ridge, a crack in the Earth's crust that runs the whole length of the North and South Atlantic (Figure 34.6(a)). It was known in the 19th Century that it existed and it was accurately mapped in the 1950s. A typical cross section across the ridge is shown in Figure 34.6(b).

This process in mid-Atlantic, whereby new material wells up to fill the gap made when sections of lithosphere separate, suggests that the Atlantic Ocean is widening and that Europe and North America are drifting apart. Modern measurements using satellites show that this is so, with the Atlantic Ocean widening by about 2.5 centimetres each year; while this may seem very slow the total widening since the Triassic Period (Figure 34.4) is 5,000 kilometres, approximately the present average width of the Atlantic Ocean. The fact that the seafloor is spreading was confirmed before the satellite measurements by some very interesting observations in the early 1960s. Molten, or very hot solid, material, including rock, placed in a magnetic field does not become magnetised. If it cools below a critical temperature, known as the *Curie temperature*, it will become magnetised — like the magnetisation of a piece of iron — and if it remains cool, thereafter it will retain the state of magnetisation it had when it reached the Curie point. In the case of rock, its magnetisation is weak, but is strong enough to be measured easily. The magma welling up to fill the gaps made by the opening crust quickly cools, and does so in the magnetic field of the Earth. We can think of the cooled magma as a magnet that is lined up with the direction of the Earth's magnetic field at the time the magma cooled below the Curie point. By dragging a magnetometer under water across the Atlantic ridge, one can record the magnetic-field direction of the Earth at the time the rocks cooled. The results obtained from this experiment are indicated in a schematic, and idealised, form in Figure 34.7. What these measurements show is that the seafloor is spreading equally on the two sides of the ridge and also that the Earth's magnetic field can suddenly reverse its direction,

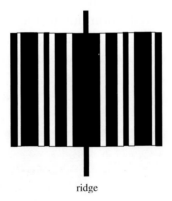

ridge

Figure 34.7 A schematic representation of magnetisation of the seafloor on either side of the mid-Atlantic ridge. Black stripes indicate magnetisation in the present direction of the Earth's magnetic field. White stripes indicate magnetisation in the opposite direction.

with the reversals happening in haphazard fashion at intervals from a fraction of a million years to several tens of millions of years.

Another kind of relative motion of parts of the crust due to mantle convection is when the lithosphere is crushed by compressive forces. When this occurs one of two things can happen. The first, illustrated in Figure 34.8(a), is that the lithosphere can be distorted and buckled both upwards and downwards to give the formation of mountains. In Figure 34.4 the path of that part of Gondwana that was destined to become India is shown and about 40 million years ago it crashed into the southern part of Laurasia. This is the event that formed the Himalaya mountain chain and the uplift of the Tibetan plateau. The movement of India northwards continues at the rate of a few metres per century and the Himalayas still rise by about 1 centimetre per year.

The second kind of outcome when crushing forces are applied to the lithosphere is that it fractures, and one side of the lithosphere slides under the other. This is illustrated in Figure 34.8(b). Such a process is called *subduction*.

Where the land is splitting and being pulled apart, as illustrated in Figure 34.5, new surface area is being created whereas the subduction process and, to a lesser extent, mountain formation both reduce the surface area. Since the surface area of the Earth remains constant (ignoring a tiny rate of reduction due to the Earth's overall cooling and shrinking) there must be a balance between the effects of rifting and subduction. The way that this occurs in the Earth was established

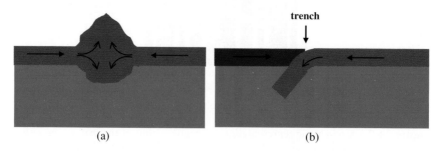

(a)   (b)

Figure 34.8   Compressive forces on the lithosphere can give either (a) mountain building or (b) subduction.

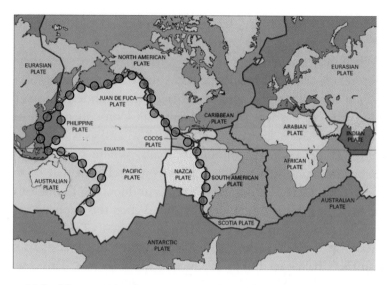

Figure 34.9   The tectonic-plate structure of the Earth's lithosphere. Red circles mark the location of the 'ring of fire' (USGS).

in the 1960s and 1970s. It was found that the lithosphere consists of a series of abutting *tectonic plates* that float and move about over the asthenosphere. This system of plates is shown in Figure 34.9. The region where plates are moving apart, as along the mid-Atlantic ridge is said to be a *divergent boundary* and if the plates are moving towards each other then we have a *convergent boundary*. For a divergent boundary only one type of process can occur; molten material from the mantle fills the breach and in the process a ridge will form. However, as indicated in Figure 34.8, for a convergent boundary there can be either mountain building or subduction and we now consider the circumstances in which each of these processes will happen.

It will be noticed that plate boundaries can sometimes be contained within continental regions, such as the northern part of the Indian Plate, sometimes be entirely within the oceans, which is true for many parts of the Pacific Plate boundary, or be at the meeting of continental and ocean regions, as occurs on the western coasts of both North and South America. An important factor that governs what happens at convergent boundaries is that continental crust has a considerably lower density than oceanic crust — 2,700 compared to 3,000 kilograms per cubic metre.

When both converging plates are continental then, since they both consist mainly of light continental rocks there is a resistance, based on buoyancy considerations, against either of them undergoing subduction into the higher-density mantle material. Consequently, mountain building occurs as is shown in Figure 34.8(a) and as is evident in the building of the Himalaya mountain chain. Where both converging plates are oceanic, with a high density comparable to that of mantle material, then there is little resistance to subduction and one or other of the plates is subducted under the other. When this happens a trench is formed, as shown in Figure 34.8(b); an example of such a feature is the Marianas Trench in the Pacific Ocean with a depth at its southern end, in the Challenger Deep, of almost 11,000 metres. This is greater than the height of Mount Everest above sea level — 8,854 metres.

The meeting of converging oceanic and continental plates, such as occurs along the western edges of the Americas, leads to the subduction of the higher-density oceanic plate, the formation of deep off-shore trenches and the formation of mountains due to the continental plate moving over the subducting oceanic plate — giving the Rocky Mountains and the Andes in the case of North and South America respectively (Figure 34.10).

Convergent and divergent motions of neighbouring plates are not the only possibilities. There is another form of relative motion of plates that has important consequences.

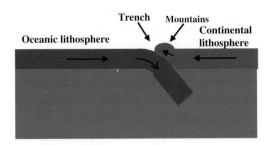

Figure 34.10   Subduction of an oceanic plate below a continental plate giving trench and mountain formation.

34.4 Volcanism

A volcano in eruption is a spectacular sight that most of those living in quieter parts of the world will never see, except on their television screens. They emit molten material coming from the mantle, usually under high pressure so that material is thrown high into the air. In extreme cases, gasses, dust and ash may be ejected into the stratosphere and then be deposited ever a wide area. An eruption of the Icelandic volcano Eyjafjallajökull in April 2010 sent ash high into the stratosphere and closed down air traffic over large parts of Europe for several days.

Volcanoes are often conical in shape, with the vent at the apex of the cone. For these *shield volcanoes* magma spills out equally in all directions, a mode of eruption that, over time, builds up the conical shape. In any part of the world, at some depth below the surface there is molten material; the conditions for this to break through to the surface to form a volcano most commonly occur in the vicinity of converging or diverging tectonic plates. For diverging plates the new oceanic crust being formed by the cooling of mantle material is thin and easily penetrated wherever pressure builds up below it. Most volcanoes produced by diverging plates are below the sea and these include black smokers, mentioned in Section 33.3 as possible sources for producing some of the molecules essential for life. There are many undersea volcanoes along the mid-Atlantic ridge; Iceland straddles this ridge and was formed by extensive volcanic outflows that eventually rose above the surface of the sea to produce an island. Within Iceland there is considerable volcanic activity, which is a source of geothermal energy of great benefit to the Icelandic community.

Volcanoes can also be produced where subduction occurs in the regions where oceanic and continental plates meet. Water carried down by the subducting plate in the region of the offshore trench lowers the melting point of the mantle material just below the lithosphere. In some circumstances, this gives molten material that can break through to the surface to form a volcano. For example, the northern boundary of the African plate runs through the Mediterranean Sea and is responsible for the volcanic activity of that region, e.g. that due to

Figure 34.11   An eruption of Mount Etna as seen from the International Space Station.

Vesuvius, Etna (Figure 34.11) and Stromboli, and also for the many earthquakes that plague that area. These conditions for volcanism to occur also exist in most of the regions abutting the Pacific Plate, giving what is referred to as the 'ring of fire' (Figure 34.9), which runs round the edge of the Pacific Ocean covering New Zealand, Indonesia, the Philippines, Japan, Korea and the west coast of the Americas from Alaska to the southern part of Chile. This region also accounts for more than 80% of the major earthquakes that occur.

Other volcanic regions of the Earth are due to 'hotspots' where the mantle is molten at comparatively shallow depths and may break through the lithosphere to form a volcano. This is thought to be due to high-temperature regions at the core–mantle boundary producing *mantle plumes*, vertical streams of very hot mantle material that penetrate close to the surface. The Hawaiian Islands in mid-Pacific are thought to have formed by volcanic activity due to this cause.

34.5 Earthquakes

Although volcanoes can be very destructive to both life and property, in general they pale into insignificance in that respect by comparison with earthquakes. To understand earthquakes we must first describe another kind of relative motion of neighbouring plates — a sliding motion in what is called a *transform boundary*. Very often this kind of motion is combined with diverging or converging plate motions. In a transform boundary the two plates have forces on them that cause a relative motion in opposite directions along a line in their common surface of contact. However, because of friction in the common surface, the two plates do not simply smoothly slide in opposite directions. Instead a strain is built up in the material of the two plates, a situation illustrated in Figure 34.12(a). As time progresses so the strain builds up until, eventually, the friction is overcome and the two plates suddenly move in opposite directions, so relieving the strain on the material, as shown in Figure 34.12(b). The energy built up over a long period of time is quickly released and the result is an earthquake. Such slippages can occur in any direction — parallel to the Earth's surface, perpendicular to it or at any angle between — and can also occur at any depth.

The strength of an earthquake is measured on the *Richter scale*, a non-linear scale that indicates the level of damage caused. An earthquake of magnitude 4.0 or less gives slight tremors that, at the higher end, can just be detected without instruments. Between 4.0 and 6.0 most buildings would be undamaged or slightly damaged at worst. From 6.0 to 8.0 will cause damage over a wide area, up to about 80 kilometres from the source for the upper end of the range. Earthquakes above 8.0 on the Richter scale are extremely destructive over many

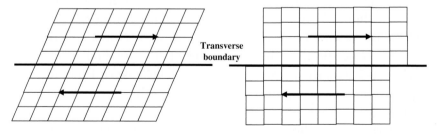

Figure 34.12   (a) Strain building up in the material on both sides of a transverse boundary. (b) The release of the strain energy leading to an earthquake.

Figure 34.13   The aftermath of the 1906 San Francisco earthquake.

hundreds of kilometres. There have been many devastating earthquakes in recorded history; a well-known one is that which struck San Francisco in 1906 (Figure 34.13). It was magnitude 7.9 on the Richter scale, quite massive, but the death toll of 3,000 was small compared to those exacted by many other strong earthquakes; for comparison, the death toll for the Tangshan earthquake, magnitude 7.5, in northern China in 1976 was well over 200,000.

The western coast of North America, stretching down as far as the northern part of California, is very unstable in the regions where the Juan de Fuca plate slides under the North American plate. California is subjected to minor earthquakes on a daily basis but it is known that every few hundred years a huge earthquake, measuring 8.0 to 9.0 on the Richter scale, occurs somewhere in western North America. To put this in perspective, the magnitude of the San Francisco earthquake was 7.9 on the Richter scale and each increase of 1.0 in Richter-scale magnitude corresponds to an increase in the energy of the event by a factor of about 30.

Apart from the direct damage due to an earthquake, another catastrophic phenomenon that often accompanies an earthquake is a *tsunami*, a tidal wave triggered by large sudden motions of the ocean

Figure 34.14   A village in Sumatra devastated by the 2004 Indian Ocean tsunami.

floor. A tsunami can travel thousands of miles across an ocean and then, when it reaches a shore, can inundate the land with a tidal wave tens of metres high. In 2004, a tsunami activated by an earthquake close to the Indonesian island of Sumatra led to 230,000 deaths, causing major damage and loss of life as far away as the Indian sub-continent and even caused minor damage on the east coast of the African continent, more than 4,000 miles from the original earth-quake. Figure 34.14 shows the wreckage of a village in Sumatra caused by this tsunami. A similar tragedy occurred in the Mediterranean region in 1908 when an earthquake in the Straits of Messina in Italy produced a tsunami that led to about 200,000 deaths.

The strongest earthquake ever to hit Japan, of magnitude 9.0, occurred on 11th March 2011. The epicentre of the earthquake was in the sea about 70 kilometres from the north-east coast of Honshu Island, at a depth of 32 kilometres. Apart from the direct effect of the earthquake itself, there was an enormous tsunami which did tremen-dous damage. In particular, a tsunami wave, some 15 metres high, overwhelmed the nuclear energy plant at Fukushima, putting safety equipment out of action, which led to overheating of some of the reactor cores and a large escape of radioactive materials. The scale of this nuclear incident was second only to the catastrophic Chernobyl reactor disaster in the Ukraine in 1986. Fortunately, on the scale of the tsunami death tolls reported for Sumatra and the Straits of Messina, the death toll was much smaller, 22,600, but there was considerable damage to the Japanese economy.

Chapter 35

Oxygen, Ozone and Life

35.1 The Role of Oxygen and Ozone

The atmosphere of the Earth consists mainly of nitrogen (78%), oxygen (21%) and argon (1%) with traces of other gasses such as carbon dioxide and methane. Nitrogen, as the main atmospheric component of a solar-system body, is found on Saturn's satellite Titan and Neptune's satellite Triton. A trace component on Earth, carbon dioxide, is the dominant atmospheric component of both Venus and Mars while methane, another Earth trace component, is about 1.4% of the atmosphere of Titan. We can see that what makes the atmosphere of the Earth unique in the Solar System is the large component of oxygen, without which there would be no advanced animal life on Earth.

All animal species depend on the availability of oxygen, either in the atmosphere or dissolved in water, since it is the element that chemically combines with carbon in food to produce the energy necessary to life, giving carbon dioxide as a waste product. The way that oxygen enters the body and carbon dioxide leaves the body is through the agency of a remarkable protein called *haemoglobin* (Section 32.2), which moves around the body within the blood stream. Haemoglobin is a large protein molecule consisting of four similar units. Because each of the units is so large, very small changes in the relative positions of neighbouring atoms, which are barely noticeable locally, have a comparatively large effect in changing the overall shape of the unit. A unit can exist in two primary shapes, in one of which there is a location on its surface that can hold oxygen but cannot hold carbon dioxide (shape A) and in the other of which the same location can hold carbon dioxide but cannot hold oxygen (shape B). The process of respiration operates as follows. Haemoglobin in blood flowing through lung tissue takes up shape A and picks up oxygen that has been drawn into the lungs by inhaling. This moves through the blood stream to the muscles of the body, in which energy is generated by the oxidation of carbon compounds. When it reaches a muscle the haemoglobin takes up shape B, at which point it deposits oxygen in the muscle tissue and takes up the carbon dioxide generated by the muscle. When the blood returns to the lung, the haemoglobin changes to shape A, releasing the carbon dioxide that is then exhaled. When the animal breathes in, the haemoglobin, now in shape A, takes up oxygen and the cycle begins again. There are many different kinds of haemoglobin, depending on the creature that possesses it and its oxygen requirements. Thus seals and other marine mammals have haemoglobin that carries much more oxygen than does the human variety. Some types of whale can remain under water for up to 30 minutes before needing to come to the surface to exhale carbon dioxide and to replenish their haemoglobin with oxygen.

Plants have a very different dependency on atmospheric gasses. Their dependency on carbon dioxide to make the cellulose and other materials they need to form their structures was described in Section 30.2.3. The energy to perform the necessary chemistry is provided by radiation from the Sun and is made possible by *chlorophyll*, which exists in green plants and enables *photosynthesis* to take place. The photosynthesis process takes in carbon dioxide from the atmosphere and combines it with water to produce the cellulose it needs and also oxygen, which the plant then releases back into the atmosphere. Animals and plants have a perfect symbiotic relationship; to animals oxygen is essential for life and carbon dioxide is a waste product whereas for plants it is the other way round.

The early Earth would have been a very hostile environment to life of any form. Animals would not have had the oxygen they needed and even plants, with plenty of carbon dioxide around, would not have been able to survive because of the intense radiation coming from the Sun — not just the sunlight for photosynthesis but harsh energetic ultraviolet radiation that rips DNA and other large molecules to shreds. The presence of life on Earth depends on it being shielded from the majority of these disruptive radiations. Curiously, it is oxygen that now protects us, but oxygen in a different from that we breathe. Oxygen in the lower atmosphere is a molecule consisting of two oxygen atoms joined together. At a height of between 60 and 90 kilometres above the Earth's surface there exists the *ozone layer*, where ozone is another kind of oxygen compound in which three oxygen atoms are joined together in the form of a triangle. The two kinds of oxygen molecule are illustrated in Figure. 35.1. Ozone is an extremely powerful absorber of ultraviolet radiation and without the ozone layer we could not survive. However, at ground level ozone is a pollutant and has harmful effects on the respiratory systems of animals, including man. It can be produced by various industrial processes and is present in the exhaust gasses from cars. It is a very powerful oxidising agent and many polymers, including rubber, will degrade if exposed to ozone. It is ironic that, especially in the 19th and 20th Centuries, seaside air was thought to be rich in ozone and beneficial to the health of those that breathed it. In the stratosphere it is, of course, beneficial to health because of the protection it provides against harmful radiation. In the 1970s it was found that a hole had appeared in the ozone layer near the South Pole and concern was expressed that some chemical agents, particularly chemicals called CFCs (chlorofluorocarbons) used in refrigerators, were penetrating into the stratosphere and breaking down the ozone layer. This led to worldwide legislation prohibiting the use of CFCs.

(a)   (b)

Figure 35.1   (a) Normal atmospheric oxygen, O_2 and (b) ozone, O_3.

35.2 The First Free Oxygen is Produced

Under the harsh anaerobic and high-temperature conditions of the early Earth, the only life form that could have evolved is archaea, but then only in environments protected from the Sun's harmful radiations — under water or in sheltered crevices. When the conditions improved then some types of anaerobic bacteria could have evolved, either from scratch by whatever process produced life from non-living material, or from some ancestor that also gave rise to archaea. One such early form of anaerobic bacterial life on Earth was cyanobacteria (green-blue bacteria or blue-green algae), which are single-celled bacteria existing in a wide range of habitats including oceans and fresh water. There are some possible fossil records of these organisms going back to 3.8 billion years before the present (Figure 33.2). They could have evolved and survived in places shielded from the very high-energy ultraviolet radiation that fell on the unprotected Earth, for example, in water and in damp crevasses. Cyanobacteria play a vital role in the ability of some types of plant to survive. They exist in symbiotic relationship with these plants, one of which is rice, the staple food of many communities in Asia. Their contribution to this relationship is their ability to fix nitrogen — that is to utilise atmospheric nitrogen to produce ammonia and oxides of nitrogen that are a food for the host plant. These bacterial organisms, like plants, can also perform photosynthesis and so convert carbon dioxide and water into cellulose and oxygen. It is generally believed that it was through the photosynthetic activity of ancient cyanobacteria that oxygen first became an atmospheric component.

The fact that the Earth was originally unprotected from very energetic radiation because of a lack of oxygen in the form of ozone gave, paradoxically, another mechanism for oxygen formation. In the early Earth, with its heavy blanket of greenhouse gasses, temperatures were much higher than they are now and water vapour would have been a significant component of the atmosphere. In Section 15.3, as part of the explanation for the high D/H ratio in Venus, we described the action of ultraviolet radiation on water, H_2O, to give a hydroxyl ion, OH, plus a hydrogen atom, H. In addition, in the early Earth carbon dioxide, CO_2, would have been dissociated to give carbon monoxide, CO, plus an oxygen atom. Individual oxygen atoms and hydroxyl ions are very reactive, and oxygen atoms would combine to give stable oxygen molecules, O_2 while oxygen, O, and hydroxyl, OH, would combine to give a stable oxygen molecule, O_2, plus a hydrogen atom, H. Because it is such a light gas, hydrogen would be lost from the atmosphere on a geologically short timescale. Other reactions would also be taking place involving methane, CH_4, and ammonia, NH_3, which, when irradiated, combine to form complex molecules some of which would be liquids or solids and thus cease to be atmospheric components.

Initially most of the oxygen being produced would not have remained in the atmosphere but would have combined chemically with various constituents of rocks. For example, in early rocks containing iron compounds, the iron was in a form that could readily take up more oxygen. There are three oxides of iron (chemical symbol Fe), *ferrous oxide*, FeO, *magnetite*, Fe_3O_4, and *ferric oxide*, Fe_2O_3, showing an increasing amount of oxygen associated with the iron. Again sulphur is a very abundant element on Earth and the effluent of volcanoes is rich in sulphur. This also provided a sink for any oxygen that was around, oxidising the sulphur to sulphur dioxide. Eventually the various oxygen-absorbing sources would have become saturated and thereafter oxygen would have begun to accumulate in the atmosphere. Some of this oxygen would have become part of the very tenuous stratosphere at great heights above the Earth where under the action of solar radiation it would have been converted into ozone — three molecules of oxygen, O_2, would have become two molecules of ozone, O_3. Once the ozone protection layer became thick enough to filter out the bulk of the harmful radiation from the Sun then life could begin to spread over the land.

The time taken for an appreciable amount of oxygen to be produced, sufficient to both set up a protective layer and to give the possibility of respiration to early life forms, was of the order one thousand million years — a long time but still only 18% or so of the age of the Earth.

Chapter 36

The Evolution of Life — From Archaea to Early Mammals

With the background of how the atmosphere changed its nature over time, we can now follow the way that life evolved, from archaea and bacteria, including cyanobacteria, to mammals, including *homo sapiens*. The time from the origin of the Earth to the present time is divided up in various ways, and here we shall consider three types of division — the *era*, *period* and *epoch*. An era is a length of time within which changes of dominant life forms of a very general kind occur — from dinosaurs to mammals, for example. Eras are further divided into periods, within which changes of life forms are considered in a more detailed way. As we approach the present time, periods are further subdivided into epochs, comparatively short intervals of time in which events take place that are closely related to the forms of life on Earth now. However, it must be understood that most changes were gradual, through the slow processes of evolution, and so the division into eras, periods and epochs is somewhat artificial — although, occasionally, there were quite rapid changes due to catastrophes that led to large-scale extinctions of life, or the effect of a mutation that gave rise to some new form of life. In indicating time, in what follows the abbreviation 'My' means 'million years' and 'BP' means 'before the present time'.

413

36.1 The Hadean Era (4,500–3,800 My BP)

This era began with the formation of the Earth, when it was an incandescent ball of silicates and iron, and concluded when it settled down with a solid crust. The temperature gradually fell to a level that enabled liquid water to form oceans, and hence provided a medium within which the necessary chemistry could take place for primitive life eventually to evolve. Continents and tectonic plates formed and, with a high temperature within the mantle, strong convection currents would have driven continental drift at a much higher rate than at present. A thick atmosphere, mostly consisting of carbon dioxide but with some nitrogen, would have been present; any gasses of lower molecular weight would probably have been lost due to the high prevailing temperature in the early part of this era. Although there were no cyanobacteria present to produce oxygen, the dissociation of water to give free oxygen, as described in Section 35.2, was releasing some oxygen into the atmosphere, which was probably beginning to produce a tenuous ozone layer.

36.2 The Archaean Era (3,800–2,500 My BP)

In this era the temperature on Earth is thought to have been similar to that at present, even though the Sun had only about 75% of its present luminosity. There are several different factors that could have compensated for the low luminosity of the Sun. The Earth still had considerable residual heat from its formation, with molten mantle material close to the surface, so leading to extensive volcanism. There would also have been a sizable heating contribution from radioactive materials, mostly contained in the Earth's crust. The main radioactive sources in the Earth are uranium-235 (half-life 4.47×10^9 years), uranium-238 (half-life 7.04×10^8 years), thorium-232 (half-life 1.4×10^{10} years) and potassium-40 (half-life 1.25×10^{10} years). From the amounts of these radioactive materials now present, it is estimated that heat coming through the Earth's surface in this era would have been almost three times the current value. Other factors compensating for the Sun's lower luminosity would have been a large greenhouse effect, due to the carbon-dioxide-rich atmosphere, and possibly less cloud cover, leading to a lower *albedo*.[a] In this era the first life appeared, firstly in the form of archaea but later as cyanobacteria, so beginning the generation of oxygen (Section 35.2) from that source.

36.3 The Proterozoic Era (2,500–543 My BP)

For most of the Archaean Era the oxygen that was being produced was largely being soaked up in oxidising various surface materials. At the beginning of the Proterozoic Era, oxygen began to accumulate in the atmosphere, initially forming about 3% of the atmosphere and steadily increasing. Even this small amount of oxygen was enabling the first forms of eukaryota to come into existence.

In this era, multi-celled organisms occurred for the first time — including multi-cellular algae and soft-bodied worm-like organisms. Cyanobacteria were also flourishing, forming large monolayer mat-like associations, usually laid down in shallow water. These could become embedded in sediments that eventually dried out and then became compressed into sedimentary rocks by the pressure of overlying material. The cyanobacteria mats left their imprint in the rocks in the form of characteristic patterned fossils called *Stromatolites* (Figure 36.1).

36.3.1 *The Ediacaran Period* (600–543 My BP)

In this period the fossils of soft-bodied creatures appear (Figure 36.2). Some of these creatures seem unlike any that now exist but others might be precursors of modern life forms. The beginning of this period is characterised by a layer of chemically distinctive carbonates, deficient in the isotope carbon-13, sitting on top of glacial deposits. It is postulated that these glacial deposits come from a period when the whole Earth was frozen from pole to pole, which was followed by a rapid evolution of life.

---

[a] The albedo of a planet is the fraction of the solar radiation it receives that is reflected back into space. The lower the albedo, the more solar radiation reaches the surface.

Figure 36.1   *Stromatolite* from Glacier National Park, Montana.

Figure 36.2   A fossil from the Ediacaran Period (British Geological Survey).

36.4 The Paleozoic Era (543–251 My BP)

During this era there was a rapid development of life that produced most of the groups of animals and plants that have ever existed — with the exception of mammals, birds and flowering plants. There were several episodes where the conditions led to the loss or decimation of many species and it ended with a major extinction of species that heralded a watershed in the types of creatures that dominated the Earth.

36.4.1 *The Cambrian Period* (543–488 My BP)

In the more than three billion years from the beginning of the Archaean Era, when life first formed, to the beginning of the Cambrian Period, living organisms had remained very simple. Now, in the 55 million years of the Cambrian Period, life became more varied, more abundant and considerably more complex — so much so that the period is sometimes called the *Cambrian Explosion*. Many new types of life evolved, including, for the first time, creatures that preyed on other organisms for food, rather than by either using decaying organic material as food or developing a symbiotic association with photosynthesising algae. A possible reason for the 'explosion' may have been that the increase of oxygen in the atmosphere allowed a higher metabolic rate that could support larger and more complex life forms.

Most life existed in the sea in this period — and at the sea bottom rather than free-floating as are most present-day fish. Creatures with hard body parts such as teeth and exoskeletons evolved, which resulted in a substantial fossil record. Figure 36.3(a) shows a fossil of a *trilobite*, an invertebrate creature with a hard shell. Trilobites were *arthropods*, creatures with an exoskeleton, segmented bodies and with limbs and other appendages that were jointed. They were notable in having primitive eyes that enabled them to sense the environment without direct contact or picking up vibrations in the surrounding water. They seem to have been the dominant life form of the period but many others existed, including sponges, molluscs and several varieties of starfish-like creatures, classified as *echinoderms*.

There were no animals with backbones in this period but there was a worm-like creature with fins, *Pikaia* (Figure 36.3(b)), which was a *chordate* with the characteristic that it had nerve fibres running along its length that connected the brain to various organs of the body — a construction that could eventually evolve into a backbone.

Figure 36.3   A selection of creatures from the Cambrian Period: (a) A *trilobite* fossil; (b) A *Pikaia* fossil; (c) A drawing of *Anomalocaris*; (d) An artist's impression of *Hallucigenia*; (e) A *Wiwaxia* fossil.

The most fearsome predator of the Cambrian Period was *Anomalocaris*, some specimens of which could grow up to 2 metres in length. There are many fossils of this creature but the drawing shown in Figure 36.3(c) gives a better impression of its form. Its prey was almost all the other creatures that existed in the Cambrian Period — trilobites and various worms and molluscs.

Another curious creature of the Cambrian Period, looking like the product of a bad dream, is the aptly named *Hallucigenia* (Figure 36.3(d)), which walked on seven pairs of spiky legs and had a back with two rows of spines that would protect it from attack from above. Since most creatures in this period lived on the sea floor many of them had spiny or bony protection on their top surfaces but were soft and unprotected on their undersides. A typical example of this kind is *Wiwaxia*, which resembled a rugby ball with spikes covering the top half (Figure 36.3(e)). *Trilobites* developed a strategy of burrowing under the sea floor so that they could attack *Wiwaxia* and other smaller creatures from underneath where they were vulnerable.

Although the Cambrian Period produced a great variety of new, larger and more complex life forms there were four occasions during the period when glaciations and mass extinctions occurred, especially of species living in warmer conditions that were unable to adapt. Another factor during glaciations was that a great deal of water on the planet existed in the form of ice over the land and, hence, shallow seas would dry up or become too shallow for some life to continue to live in them.

By the end of the Cambrian Period there were some incursions of sea creatures onto land but there were no true flora on land, although there may have been algae and lichens in damper locations.

36.4.2 *The Ordovician Period* (488–444 My BP)

The Ordovician Period began just after a major extinction event at the end of the Cambrian Period and was itself a period in which many life forms rapidly developed. Most of this life still dwelt in the sea, including trilobites that had survived the extinction and which developed many new varieties. Also present at this time were *cephalopods*, creatures similar to the modern octopus and squid (Figure 36.4), which were the leading predators of their time. Other creatures present were sponges, corals, *gastropods* — snail-like organisms — and *crinoids*, which were echinoderms looking like feathery ferns. It is claimed that the first true vertebrates, fish, appeared in this period although it is possible that they had actually first evolved in the Cambrian Period. However, it is certain that the very first jawed fish appeared by the end of this period. Prior to that time fish were like lampreys, with suckers rather than mouths.

A completely new, and abundant, species in the Ordovician Period were *brachiopods* (Figure 36.5), creatures that resembled modern clams but that were biologically completely different.

Figure 36.4   Various forms of cephalopods in the Ordovician Period (Ernst Haeckel, *Kunstformen der Natur*, 1904).

Although faunal life at this time was confined to the sea the first flora began to colonise the land, probably exploiting tidal regions that would enable a gradual adaptation to occur. This flora resembled the present liverwort — flat, branching, ribbon-like plants.

During all the period of time from when land first formed, the solid land was moving over the fluid mantle of the Earth, as described in Chapter 34. At the end of the Ordovician Period, Gondwana had

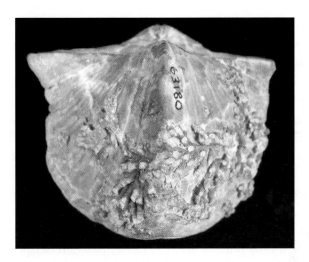

Figure 36.5   A brachiopod (*Hederella*).

established itself over the South Pole. The present continent of Antarctica contains vast quantities of water in the form of ice, some 70% of the Earth's fresh water, and this is because of the great depth of ice that can form over the solid Earth. When the whole of Gondwana was situated over the South Pole, even greater quantities of water were trapped there as ice. This led to a fall in sea level over the Earth and the draining of shallow seas. Because of this phenomenon some 70% of all types of living organisms had died by the end of the Ordovician Period.

36.4.3 *The Silurian Period* (444–416 My BP)

Despite the considerable extinction at the end of the Ordovician Period many of the creatures from that period, such as brachiopods, crinoids and *nautiloids*, types of cephalopod, survived into the Silurian Period. Several factors came into play in the Silurian Period that favoured a new flowering of life forms and a move onto land. The climate became warmer so melting the thick ice layers on Gondwana, which was still at the South Pole, and so increasing both the extent and depth of the oceans. At the same time, due to continental drift and the collision of tectonic plates, new mountains were being built so that some areas previously under water, or in coastal regions, gradually became dry land. Remember, Charles Darwin found seashells high in the Andes mountains (Section 31.4). Plant life that had adapted to tidal regions, which never completely dried out, now had to adapt to drier conditions to survive. The new land plants were mostly small, a few centimetres in height but they had rigid stems and root-like systems, the first step towards the kinds of plants on Earth today.

Another important factor that enabled a considerable colonisation of the land at this stage was that the ozone layer was well established so the danger from damaging radiation was reduced to a tolerable level. Newly-formed tidal regions acted as nurseries where adaptations could take place and where new strategies for living on land could evolve. Arthropods — insects, spiders and centipedes — moved onto land, some of them much larger than present creatures of similar kind. One of the species, *Eurypterid* (Figure 36.6), a fierce predator similar to a modern scorpion, was two to three metres long but since it could not support its full weight on land it lived in shallow marine environments.

Fish, the only vertebrate form of life, were becoming more common. Early forms of fish were jawless with backbones consisting of soft cartilage but, towards the end of this period, a few fish with jaws and hard backbones had evolved.

36.4.4 *The Devonian Period* (416–360 My BP)

The Devonian Period marks a transition from a world with flora and fauna that seem to have little connection to the present time to one with some features that would be familiar to us. During this period there were significant movements of land masses with the continent of Laurasia being formed by the collision of two smaller continents, *Euramerica* and *Baltica*. The formation of Laurasia and other movements of land led to a considerable amount of mountain building.

In the sea new species of jawed fish were present, including the armoured *Placoderms* (Figure 36.7), which had cartilage-type backbones, and ray-finned fish. Sharks developed during the Devonian

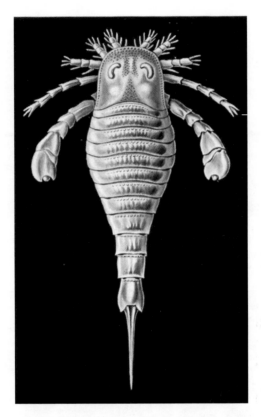

Figure 36.6 *Eurypterid* (Ernst Haeckel, *Kunstformen der Natur*, 1904).

Period, possibly evolved from placoderms by the loss of the surface armour but still retaining the cartilage backbones. Great reefs formed in the sea, similar in type to present reefs but derived from different living organisms. Many new invertebrate species also evolved, in the sea including ammonites, which could grow to a huge size and which left behind many splendid and clear fossils in Devonian rocks (Figure 36.8).

A particularly interesting Devonian fish is the *Coelacanth* (Figure 36.9), once thought to be long extinct. In 1938 a living coelacanth was caught in the Indian Ocean; since then others have been found from time to time.

Life was now well-established on land. A few fish had developed structured fins that enabled them to 'walk' over the sea bed and these

Figure 36.7   A typical *placoderm* (Tiere der Urwelt, *Creatures of the Primitive World*, 1902).

Figure 36.8   A large ammonite fossil.

evolved into the legs of the first four-legged land-walking creatures (*tetrapods*), which were amphibian. One of these amphibians, *Ichthyostega* (Figure 36.10) shows clearly the evolution from a fish-like form.

Figure 36.9 An artist's impression of a *coelacanth* (Robbie Cade).

Figure 36.10 *Ichthyostega.*

The two super-continents, Laurasia and Gondwana, spanned the equator, making conditions on land warm and conducive to the development of plant life. During the Devonian Period plants developed true root systems that were able to exploit moisture over a large area and vascular tissues to distribute it to all parts of the plant. Plants had thus become conditioned to living away from large sources of water. They had grown much bigger and spread across the land and propagated by producing seeds, rather than spores as had the first land plants. Of great significance is that the first tree, *Archaeopteris* (Figure 36.11), came into being, a tree with fern-like leaves growing to a height of 15 metres or more — one that would not be much out of place in a modern setting.

36.4.5 *The Carboniferous Period* (360–299 My BP)

During this period the super-continents Gondwana and Laurasia were moving towards each other, eventually to form the single continental land mass, Pangaea (Section 34.2). This movement was producing mountains and land uplift in some regions. The residues of crinoids, and other hard-shelled creatures, were contributing a great deal of

Figure 36.11   The first true tree, *Archaeopteris.*

limestone (calcium carbonate, $CaCO_3$). There was an explosion in both the number of species of plants and in the total quantity of those plants, which flourished in the warm, damp conditions that prevailed during most of this period. Many kinds of tree evolved and giant ferns provided the undergrowth for the resultant forests. When this vegetation died it did so under anaerobic conditions, with few bacteria that would help decomposition, and they first formed a peaty deposit that later, under pressure from overlying material, was converted into coal. The name *Carboniferous Period* derives from the large deposits of coal that originated in this period.

Brachiopods were still widespread but trilobites were dying out and were virtually extinct by the end of the period. Tiny filter-feeding animals, a few millimetres in average dimension, called *bryozoa*

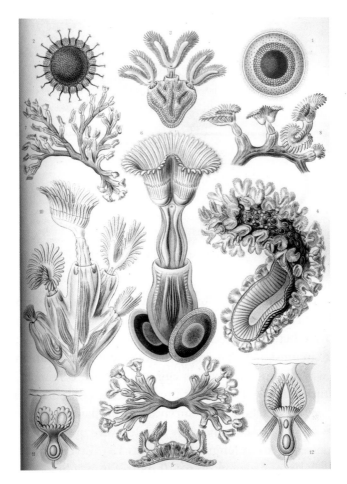

Figure 36.12   Forms of *bryozoa* (Ernst Haeckel, *Kunstformen der Natur*, 1904).

(Figure 36.12), which attached themselves to rocky surfaces in great colonies, occupied the many shallow seas that existed at this time.

The armoured placoderms that had dominated the seas in the Devonian age had become extinct and were replaced by many other species of fish, ones that looked like modern fish.

One result of the enormous expansion of plant life is that photosynthesis greatly increased the amount of oxygen in the atmosphere and this enabled the development of ever more complex creatures with

Figure 36.13   An impression of an early reptile, *Hylonomus*.

higher rates of metabolism. Reptiles developed from amphibians and colonised the land well away from large stretches of water. Their leathery skins were better adapted to living under drier conditions; another important circumstance that enabled movement away from amphibian existence was reproduction via an egg with a hard, originally leathery, outer coating, a form of shell that prevented the contents from drying out. The greater likelihood of survival of the hatchlings meant that fewer eggs had to be laid. The early reptiles were quite small, typically 20 centimetres or so in length (Figure 36.13). By contrast trees had become very large, growing up to 40 metres in height.

Insects also flourished in the warm, damp and well-oxygenated conditions. A giant precursor of the dragonfly, *Meganeura*, had a wingspan of up to 75 centimetres and there were giant millipedes, *Arthropleura*, that were one-and-a-half metres long.

36.4.6 *The Permian Period* (299–251 My BP)

By this time the continents had merged into one great land mass, Pangaea (Figure 36.14), which was drifting northwards from the equator, so leading to a cooler climate. The interior of this super-

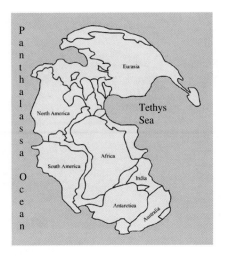

Figure 36.14   The Pangaea super-continent showing regions of the present continents (USGS).

continent was far from any ocean and was desert-like in climate — dry with hot days and cold nights.

Plants developed that were adapted to these drier conditions and which propagated by distributing seeds. Ferns and conifers became the dominant vegetation of the period.

With the drier conditions, reptiles took over from amphibians as the major land-dwelling species. The best known of these reptiles is the *Dimetrodon* (Figure 36.15). Creatures began to evolve which resembled mammals but still had reptilian characteristics; these would become the first mammals in due course — giving birth to live young (with a few exceptions) and with females having mammary glands to feed infants with milk.

At the end of the Permian Period there occurred a great extinction of species, of unknown cause, which affected more than 90% of the marine species present; corals, which had thrived during the greater part of the Permian Period, were decimated. The reptiles seemed most able to survive this extinction and they were destined to become the dominant species on land.

Figure 36.15   A lizard *Dimetrodon* from the Permian Period.

36.5 The Mesozoic Era (251–65.5 My BP)

This era began with reptiles as the dominant kind of life but by the end of the era dinosaurs had become dominant. This era also saw the arrival of flowering plants, birds and mammals. It began with a great extinction and finished with another, and even greater, extinction of species.

36.5.1 *The Triassic Period* (251–200 My BP)

The land was still in the form of the single continent Pangaea, but during this period it began to break up again into Gondwana and Laurasia until, at the end of the Triassic Period, the two continents were positioned as shown in Figure 34.4.

A very common fossil from this period is that of the *Belemnite*, a squid-like creature that differed from its modern counterpart in having an internal structure consisting of bony plates, although it was not a vertebrate since it had no backbone.

The hip structures of some reptiles were changing so that they adopted posture more upright than the spread-out posture of the original reptiles; these modified reptiles mostly died out. Reptiles also became dominant sea creatures although, like modern whales,

(a)   (b)

Figure 36.16   Triassic Period marine lizards (a) *Ichthyosaur* and (b) *Plesiosaur*.

they breathed through lungs rather than gills. They had streamlined bodies for efficient motion through water. Two impressive examples of Triassic marine reptiles are *Ichthyosaurs* (Figure 36.16(a)), some eight metres in length, and *Plesiosaurs* (Figure 36.16(b)), about 12 metres long, including its long neck. The fabled monster 'Nessie', reputed to live in Loch Ness in Scotland, is reported by those who claim to have seen it to resemble a plesiosaur. Towards the end of the Triassic Period there were some flying reptiles, *Pterosaurs*, but the general opinion is that these were not the ancestors of modern birds.

The first dinosaurs were beginning to appear towards the end of the Triassic Period. Although dinosaurs are sometimes thought to be similar to lizards — indeed the Greek word σαυρα (saura) means lizard — they are different in two important respects. Firstly, they were warm-blooded, unlike the cold-blooded lizards, and, secondly, their bone structure was different. Because of their hip and leg structures they could stand upright and they had hands that could be used for grasping objects. Dinosaurs were eventually destined to rule the Earth — but not yet.

Another significant development towards the end of the period was the appearance of reptiles that were beginning to resemble mammals, for example *Thrinaxodon* (Figure 36.17), a carnivore about 30 centimetres long, which may have been fur-covered and warm-blooded. True mammals later evolved from creatures of this kind.

Figure 36.17   The mammal-like lizard *Thrinaxodon*.

However, the cold-blooded lizards were better able to survive in the heat of Triassic deserts and the mammal-like lizards may have thrived mainly in the cooler regions of Earth at that time.

The trees and ferns which had flourished in the Carboniferous Period and provided modern coal deposits could no longer survive in the comparatively dry Triassic Period. The dominant vegetation was now conifers and ginkgos, the latter being a giant tree that still survives in some parts of eastern Asia.

36.5.2 *The Jurassic Period* (200–145 My BP)

The separation of Laurasia and Gondwana continued and the individual super-continents began to break up to form the continents we know today. The splitting of the continents, including within the Tethys Sea (Figure 34.4), would have produced abundant volcanic activity. There would also have been plate collisions producing mountains, especially in those regions along the western side of Pangaea corresponding to the western coasts of present North and South America. The climate was warmer than that at present and the dominant plant life consisted of palm-like trees, conifers, ferns and various smaller species. The warm and stable climate over the period allowed the development of many new species of plants and animals.

Life in the sea flourished in this period. *Ichthyosaurs* and *Plesiosaurs* increased in number as did belemnites and ammonites, all of which

(a)   (b)

Figure 36.18   Herbivore dinosaurs of the Jurassic Period (a) *Brachiosaurus* and
(b) *Stegosaurus*.

were efficient predators. There was an abundance of vertebrate fish,
sharks and rays.

Dinosaurs dominated the animal world. The herbivores had an
abundance of food and grew to huge sizes. The largest dinosaurs of
the period were *sauropods*, which stood on four legs, had long necks
and a counterbalancing long tail; a typical sauropod, *Brachiosaurus*
(Figure 36.18(a)), weighed up to 55 tonnes. Another, and different,
type of herbivore dinosaur was the *Ornithopod*, which had a hip structure resembling that of present-day birds. Figure 36.18(b) shows
Stegosaurus, about nine metres long with two rows of spines along its
back. It did not have the long neck that enabled brachiosaurus to feed
from high herbage on trees so it almost certainly fed from ferns and
other close-to-ground plants.

The carnivore dinosaurs of this period were also gigantic. They
walked on two legs, their front legs being more like arms, short and
with 'hands' equipped with formidable claws. A typical example of a
carnivore of this period is *Allosaurus* (Figure 36.19), some 11 metres
in length.

Many true mammals had evolved, although they were small, mouse-
like creatures. Much larger creatures were taking to the air, for example,
Archaeopteryx (Figure 36.20). Flying dinosaurs of the time, including
Archaeopteryx, are believed to be the ancestors of modern birds.

As with many periods, the end of the Jurassic Period is marked by
a minor mass extinction.

Figure 36.19   The fossilised remains of an Allosaurus (San Diego Natural History Museum).

36.5.3 *The Cretaceous Period* (145–65.5 My BP)

The division between Laurasia and Gondwana, and their subdivision into the present continents, continued, with the North Atlantic just beginning to be formed by a break up of Laurasia, and the South Atlantic by a separation of South America and Africa. The rift in the expanding Atlantic Ocean produced many undersea volcanoes and mountains. The sea level was much higher than it is now, by as much as 100 metres.

An important development during this period was the arrival of flowering plants, and also many new kinds of grasses and trees. The flowers developed a symbiotic relationship with insects that benefited from the nectar produced by the flowers while they performed the task of efficiently moving pollen from one flower to another.

New dinosaurs that appeared in this period included *Triceratops*, a large bird-hipped herbivore with three horns and a bony collar (Figure 36.21(a)), the herbivore *Hadrosaurus*, depicted in Figure 36.21(b), which was up to 10 metres long and weighted up to 7 tonnes, and the fearsome carnivore *Tyrannosaurus Rex*, shown in Figure 36.21(c), which was up to 12 metres long and 5 metres tall and was the largest carnivorous creature ever to walk the Earth.

Figure 36.20 A fossil record of the flying dinosaur *Archaeopteryx* (Muséum national d'Histoire naturelle, Paris).

Figure 36.21 Three dinosaurs of the Cretaceous Period: (a) *Triceratops*, (b) *Hadrosaurus* and (c) *Tyrannosaurus Rex*.

Mammals were also on the increase in this period, but they were an insignificant part of the fauna. A new feature of mammals was the evolution of the placenta and the development of offspring within the female's body.

At the end of the Cretaceous Period a mass extinction (the KT event) occurred that wiped out all larger forms of life, including the dinosaurs, although smaller mammals largely survived. A probable cause of this extinction (disputed by some scientists) was the impact on Earth of a large asteroid, 10 kilometres in diameter; the energy released by such an event would have been the equivalent of exploding 100,000 large hydrogen bombs. The evidence for this event is a thin deposit, rich in the element iridium and found throughout the world, with an age of about 65 million years. Iridium is comparatively rare on Earth but is more common in some meteorites, and hence, presumably, in asteroids. There is also evidence for a crater that might have been produced by such an event in the Yucatan Peninsula in Mexico. How this catastrophe led to extinction is a matter of conjecture. Apart from the obvious direct effects of the blast there could have been massive tsunamis, and dust thrown high into the atmosphere could have blotted out the Sun for many years, so killing off much of the Earth's vegetation. Without vegetation herbivores cannot live, and without herbivores carnivores cannot live.

The KT event not only ended the Cretaceous Period (the K in KT is from the German name for this period — *Kreidezeit*) but also heralded the beginning of the Tertiary Period, the first period of the next era. Dinosaurs were no more, but mammals were there in the background, ready to take over the domination of life on Earth.

Chapter 37

Early Mammals to Man

While dinosaurs dominated the Earth there was no opportunity for any other type of creature to evolve to a size that could challenge them. Early mammals had developed, but their defence was that they were small and inconspicuous and that they did not compete with dinosaurs in any direct way. Now that dinosaurs had vanished, the field was clear for mammals to evolve and to become dominant.

37.1 The Cenozoic Era (65.5 My BP to Present)

This era is divided into two periods, the *Tertiary* and *Quaternary*, which are each further subdivided into shorter-interval epochs that enable us to follow evolutionary processes in greater detail as we approach the present.

37.1.1 *The Tertiary Period* (65.5–1.8 My BP)

This period saw the rise of mammals as the dominant form of fauna. At the beginning of the period the first primates appeared; they were the first step on the evolutionary ladder that would eventually give modern man. Although very different from man, these early primates had some human characteristics. They had five digits on their fore and hind limbs, with opposable thumbs enabling them to grasp and manipulate objects. For most primates these digits had nails, rather than claws that were more suitable for digging into surfaces, be they the trunks of trees or the flesh of prey. The eyes of most primates give good colour vision and tend to be at the front of the face, so giving highly overlapped fields of view from the two eyes, and, hence, well-developed stereoscopic vision. Many, but not all, primates can hold their bodies upright and walk on two legs — which frees their hands for other activities while they walk. Another common characteristic is that, relative to other creatures, they have a large brain relative to their size.

The period ended with the evolution of hominids, mammals that were definitely man-like in structure and appearance, and it is likely that the first predecessor of modern man had appeared on the scene.

37.1.1.1 *The Palaeocene Epoch* (65.5–56 My BP)

The continents had now moved close to their present positions. However, Europe and North America were still partially joined, Australia was attached to Antarctica and India was still on its journey towards Asia.

The only remnants of the dinosaurs were birds that, by their very nature as flying creatures, could not be physically very large. Mammals, both herbivores and carnivores, were increasing in size, although still small by modern standards. The mammals of this period were fairly primitive compared with modern mammals and they had not developed the specialisations that make modern mammals so successful in survival terms. Some of the mammals still laid eggs, as does the modern platypus, and there were many marsupials, mammals with pouches like kangaroos.

Forests of broad-leafed trees covered large parts of the Earth and provided a habitat for many early mammals. There is some fossil evidence that a squirrel-like arboreal creature, which may have been a precursor of primates, evolved during this time.

37.1.1.2 *The Eocene Epoch* (56–34 My BP)

Europe and North America had now clearly separated and the Atlantic Ocean had completely formed, from pole to pole. The boundary

Figure 37.1 *Hyracotherium.*

between the Palaeocene and Eocene Epochs was notable for temperatures so high that tropical rainforests could exist at the poles. However, by the end of the Eocene Epoch the Antarctic icecap had formed and has been in place ever since.

Hyracotherium (Figure 37.1), the forebear of horses and similar creatures, came on the scene, although they were small — about the size of an average present-day dog. Over time these creatures evolved to become much larger. Horses, zebras and the rhinoceros are among the modern survivors of this branch of evolution.

In this epoch some mammals reverted from living on land to living in the sea, a complete reversal of the evolutionary development that had established life on land. The ancestors of whales appeared, although they did not resemble the whales we know today. An intermediate stage in this evolutionary process is represented by *Ambulocetus* (Figure 37.2), a four-legged mammal, about three metres in length, which spent a great deal of its time in water. It had acquired some of the characteristics of a modern whale — it had no external ears and had an adaptation of its nose that enabled it to swallow under water. It was a carnivore that probably hunted much as a crocodile hunts today, by hiding under water and grabbing at its prey.

An important development in this period was the appearance of the first primates, which resembled modern *prosimians* such as

Figure 37.2   *Ambulocetus*, an ancestor of the whale.

Figure 37.3   Lemurs.

tarsiers, bushbabies and lemurs (Figure 37.3). A prosimian is a primate that is neither a monkey nor an ape. These early creatures had many of the characteristics of primates; with their five-fingered hands they could manipulate objects and climb efficiently and their visual systems gave improving colour discrimination and stereoscopic vision, the latter being helped by a smaller snout that gave better overlap of the visual fields of the two eyes.

For a primate to take an upright posture, as distinct from walking on four legs, the hole in the skull where the spinal cord enters the head, the *foramen magnum*, must move from the back of the skull towards the centre. This process was happening during the Eocene Epoch, suggesting that, like modern lemurs, when sitting or moving about, the prosimians of the time had their bodies in an erect position.

During the Eocene Epoch, prosimians were present in all parts of the world that had forests and a suitably high temperature. However, by the end of the epoch the first monkeys were appearing, which occupied a similar habitat, and most species of prosimians became extinct because they could not compete with the new arrivals. Only on the island of Madagascar, where monkeys and apes did not evolve, were the prosimians able to flourish, and lemurs still exist there in great numbers. In other locations lemurs have become nocturnal creatures, keeping them clear of the larger primates.

37.1.1.3 *The Oligocene Epoch* (34–23 My BP)

Within this epoch the Indian Plate crashed into the southern flank of Laurasia (Figure 34.4) and Antarctica, which had separated from Australia, had reached the South Pole and was becoming ice covered. The overall world climate changed from very wet and tropical to somewhat drier and sub-tropical, and grasses developed that produced huge savannah regions.

The change in conditions favoured the domination of mammals such as deer, horses, cats and dogs, and the first elephants appeared. The largest mammal from this epoch was *Paraceratherium*, a giant

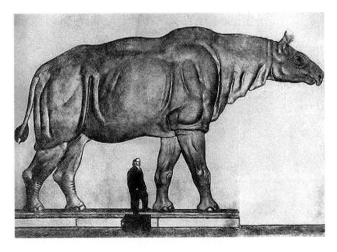

Figure 37.4 A drawing of *Paraceratherium*, compared to a man.

hornless precursor of the rhinoceros, which was more than 6 metres tall (Figure 37.4). The formation of the Antarctic ice sheets caused the seas to retreat and there was a loss of many marine species, including the first primitive whales, which were later replaced by modern whales.

The prosimians evolved into the first *Old World monkeys*, which occupied tropical and grassland regions of Africa and Asia. They were mostly small, although some approached the size of a small dog. A type of Old World monkey is the *Talapoin* shown in Figure 37.5. It is easy to imagine a slow evolutionary change from a lemur to this creature; the snout has become smaller and it has fewer teeth, eyes more forward looking and a larger brain. It weighs about 1 kilogram and is tree-dwelling. At the other extremity of Old World monkeys the modern *Mandrill* lives on the ground and can weigh up to 50 kilograms.

In the early part of the epoch, about 30 My BP, a new group of monkeys appeared in South America, the *New World monkeys*. These are thought to have derived from earlier Old World monkeys in Africa that migrated across the Atlantic Ocean on floating islands of vegetation and soil that were torn off the land by violent storms. However, by whatever means they reached South America, they evolved into new, distinctive forms. They have flatter noses, with nostrils at the side, and they have long prehensile tails, useful for holding on to

Figure 37.5   *Talapoin*, an Old World monkey (Friedrich Wilhelm Kuhnert).

branches — something the shorter-tailed Old World monkeys cannot do. Another interesting difference is that all the New World monkeys, with the exception of *howler monkeys*, lack full-colour vision. They only have two visual pigments in their retinal photoreceptors, rather than the three for Old World monkeys, and so they have a more restricted range of colour discrimination.

37.1.1.4 *The Miocene Epoch* (23–5.3 My BP)

Continental drift continued and the arrangement of land was recognisably similar to that at present, except that North and South America were still separate. The Arctic icecap had formed, but, since there was no land beneath it and the ice was floating, this had no implications for sea levels. Mountain ranges were building in the western Americas and northern India. The climatic trend was towards being cooler and drier. Extensive grasslands provided a bountiful environment for herbivores, including ruminants. The fauna for the most part consisted of modern species, for example, wolves, horses, camels, deer, crows, ducks, otters and whales.

During this epoch the first apes evolved from monkeys. They are distinguished from monkeys by not having tails, by usually being much larger and by having a more human appearance. The family of *hominoidae* or *great apes* includes chimpanzees, orang-utans, gorillas, bonobos (a dwarf chimpanzee) and humans. Like their monkey forebears they are mostly tree-dwelling — the exceptions being the gorilla and humans. There is also a family of *lesser apes*, which includes gibbons of various kinds. One of the early apes was *Proconsul* (Figure 37.6), a tree-living ape that occupied the forests of Africa between 21 and 14 My BP.

During this epoch about 100 different kinds of ape evolved, one or more of which were precursors of *hominids*, the ancestors of modern man. They lived mainly in Africa and in southern grassland areas of Europe. However, colder conditions in Europe towards the end of the Miocene Epoch led to the extinction of many species of apes; the remainder survived in the more benign and warmer conditions in Africa and by migration to Southern Asia.

Figure 37.6   An early ape — *Proconsul.*

37.1.1.5 *The Pliocene Epoch* (5.3–1.8 My BP)

The final step in creating the modern arrangement of continents took place when the land bridge formed between North and South America, an event that allowed the mixing of the species in the two continents. Ice sheets existed at both poles and the Earth continued to cool.

There was a steady evolution of primates, with hominids, close relatives of man in direct descent, having prominent jaws and larger brains, appearing towards the end of this epoch. The oldest fossil record of a hominid dates from just over 5 million years ago and was discovered in Ethiopia. A younger, almost complete, skeleton of a female hominid, called Lucy by its discoverers and designated as of species *australopithecus africanus,* dates from about 3 My BP and may represent the stage at which chimpanzees and humans diverged in their evolution. It is notable that humans and chimpanzees have 98.9% of their coding-DNA in common. Lucy was 1.1 metres tall and weighed about 29 kilograms. From her pelvic bones it can be deduced that she walked upright on two legs. An artist's impression of the hominid *australopithecus afarensis,* earlier than Lucy, as deduced from its fossil remains, is shown in Figure 37.7.

An early hominid that appears at the end of the Pliocene Epoch, about 2.2 My BP, and that recognisably spans the appearance gap

Figure 37.7 *Australopithecus afarensis* (CosmoCaixa, Barcelona).

between *australopithecus africanus* and *homo sapiens* is *homo habilis* (Figure 37.8(a)). *Homo habilis* had a brain size in the range 500–800 cubic centimetres, greater than that of *australopithecus africanus*, which was about 450 cubic centimetres, and a part of the brain associated with speech had developed somewhat so he may have been capable of elementary speech communication. There is evidence that he was making and using primitive stone tools. He was of similar size to

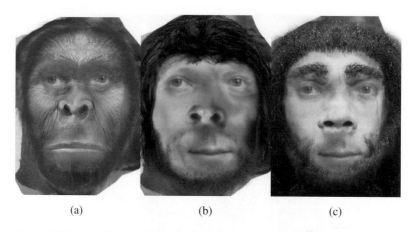

Figure 37.8   (a) *Homo habilis,* (b) *Homo erectus* and (c) *Homo Heidelbergenss.*

australopithecus africanus, with average height 1.5 metres and weight 45 kilograms — which is also similar to that of modern pygmy people in the Congo region of Africa.

The end of the Tertiary Period marks the beginning of the evolutionary process that leads to modern man — *homo sapiens* — all happening in a space of about 3 million years. Seeing the changes in that period, from the creature portrayed in Figure 37.7, probably not very different from Lucy, to man as seen today, gives a better understanding of how changes wrought over 4,000 million years could give present life forms, starting with the simple bacterium.

37.1.2 *The Quaternary Period* (1.8 My BP to present)

This is a period of very rapid change, not so much in the rate of evolution, which had been even faster in the past, but in the rate of development of the capability of the species that evolved, by which we mean hominid development leading to mankind. From the time of australopithecus africanus, various strands of development led towards ever more advanced forms of hominid in terms of brain size and concomitant ability to manipulate nature to their own ends. Many of these evolutionary strands separated by migration into different environments and evolved separately to become distinctive in both appearance and lifestyle. Most of these strands became extinct, although their fossil remains are frequently uncovered. Thus *homo antecessor*, the fossil remains of which were discovered in Spain in 1977, is dated to 780,000 years ago and shows a mixture of characteristics of modern man and of more primitive hominids. There is even doubt that the fossil remains are all of one species but, in any case, it is difficult to link *homo antecessor* into the line of development of modern man. Here we shall concentrate on three kinds of hominid that seem to indicate a sequence of development between *homo habilis*, the first primate distinguished by the term 'homo', indicating 'man', and modern man, described as *homo sapiens sapiens* — 'wise wise man'.

37.1.2.1 *The Pleistocene Epoch* (1.8 My to 11,500 years BP)

The climate of this epoch was punctuated by repeated glacial cycles that, at their peak, covered 30% of the Earth's surface. At maximum glaciation a great deal of the Earth's water was in the form of ice and the seas retreated. In warmer periods the seas advanced again. Because of the decrease in liquid water available to be evaporated, the climate became drier and extensive deserts formed.

Species of animals that became extinct by the end of this period include mammoths, mastodons (like woolly elephants), sabre-toothed tigers, glyptodons (similar to armadillos and as big as a small car) and ground sloths.

At the beginning of the epoch, about 1.8 My BP, a new kind of hominid, *homo erectus,* appeared on the scene. From fossil skulls it is deduced that he was beginning to resemble modern man (Figure 37.8(b)). His brain size was between 750 and 1,225 cubic centimetres, a considerable advance on *homo habilis,* his stone tools were much more sophisticated than those of *homo habilis* and there is some evidence that he had mastered the use of fire. *Homo erectus* was found in many locations in the world, in Africa, Asia and Europe, with slightly different physical characteristics in different locations. The overall impression from the various fossil remains is that they were of similar stature to modern man but much more robust.

The next stage in the evolutionary pathway to modern man is an early form of *homo sapiens, homo Heidelbergensis* (Figure 37.8(c)), which was certainly present 500,000 years ago but may have been slightly earlier. This species of hominid bridges the gap between *homo erectus* and modern man. The brain size averaged 1,200 cubic centimetres, but was larger for some individuals, and the skull shape was closer to that of modern man, although still with prominent brow ridges and a receding forehead and chin. They were skilful stone-tool makers and had mastered the use of fire.

Humans essentially attained their present form during this period. There were two lines of human development, *Neanderthals,* who developed in Europe and western Asia about 230,000 years ago, and *homo sapiens,* who first evolved in Africa. Neanderthals were shorter and more heavily built than *homo sapiens,* and were well adapted to living in a cold climate. They appeared to have larger brains, although, since they were heavier than modern man, the ratio of brain size to weight was probably similar. When the skull and other bones of Neanderthal man were first discovered in the early 19th Century, images of their postulated appearance were such that they would not have seemed out of place in a zoo. While the shape of their skulls was different from that of modern man — they had a definite bulge at the back, were distinctly beetle-browed and had flatter noses that suited living in a cold climate — the dominant view now is that they would certainly have stood out in a modern setting, but that they were mostly similar in facial appearance to present man. An impression, based on this interpretation of their appearance, is given in Figure 37.9. Neanderthals coexisted with *homo sapiens* for some time and the cause of their extinction is not fully understood. The last traces of them were found at Gibraltar, where they existed until 24,000 years ago.

37.1.2.2 *The Holocene Epoch* (11,500 years BP to present)

This is the age of modern man (Figure 37.10). The climate has been relatively stable, and warm, during this period with minor blips such as the few hundred years of lower temperatures ending in about

Figure 37.9 An impression of Neanderthal man.

Figure 37.10 A grandfather and grandson in Papua New Guinea.

1800 AD. During the reign of Elizabeth I the River Thames occasionally froze over to the extent that the people would hold a 'frost fair' on the Thames itself. Such an event is now unimaginable.

Homo sapiens sapiens has spread over the whole habitable globe in exponentially increasing numbers and has developed technology to the point where the environment is more influenced by man than by natural events — with results as yet uncertain but a cause for concern.

Chapter 38

Man and the Earth

38.1 Environment, Chance and Evolution

The story of the evolution of the Earth and its atmosphere, and of the living organisms that inhabited it in the past and present, has involved a chain of causally-related events, initiated by the internal cooling of the Earth that precipitated motions of its solid surface, and the fortuitous (or inevitable?) beginning of life. The heating and cooling of the Earth's climate and the rise and fall of oceans, which changed the conditions under which life had to survive, combined with the principles of Darwinian theory — the survival of the fittest — not only modified the characteristics of species but also gave rise to completely new species. At the root of this story is the role of DNA, unknown to Darwin but driving the whole process of evolution. The effects of energetic solar radiation, and of some chemical agents, occasionally produce a change in DNA structure, a mutation that can radically alter the nature of the affected organism and its ability to survive in the prevailing conditions. The vast majority of such changes are harmful, so that the organism is adversely affected and the mutation, either slowly or quickly, is eliminated from the population. It can be thought of as a failed experiment. However, once in a while the experiment succeeds and the mutation is of benefit. Now there is a member of the population of organisms that has an edge in the battle for survival. It passes on the mutated gene and soon there is a small colony of advantaged individuals within the population. The same mathematics that gave Figure 31.9 shows that, within a time that is short by geological standards, the mutation can completely take over and a new kind of organism has replaced the original one.

The process that has just been described is one dictated by chance. The mutation that succeeded happened because a particular change happened to a particular part of the individual's DNA. Again, a mutation that might have been beneficial in one climatic environment could be harmful in another; a mutation that tends to thicken the fur over an animal's body will be a good thing in an ice age but bad when the conditions are tropical. The whole process of survival of a particular species of organism greatly depends on luck — on the likelihood that when the conditions change a mutation will occur that will give an increased chance of survival under those new conditions. If so, then the species will survive, albeit in a modified form; if not, then the species will become extinct. The world is a great casino where the nature and rules of the game keep changing, where the odds are stacked against the punter and the penalty for losing is extinction. Once in a while extinctions have been on a massive scale — the whole Earth has frozen, the seas have dried up or an asteroid has fallen — but as long as some life exists in some form, somewhere, it will act as a nucleus for radical new species to evolve. If the asteroid had missed the Earth 65.5 million years ago then the dinosaurs that had ruled the world for more than 100 million years might still be ruling it today — but it did not miss. This catastrophe, which wiped out a large proportion of life on Earth, allowed mammals to evolve, including ourselves. However, it could be argued that *homo sapiens sapiens* may be just the latest of a sequence of catastrophes to threaten life on Earth!

38.2 Man Arrives and Begins to Manipulate Nature

The arrival of mankind as a species of life introduced a new factor into the evolutionary game. For the first time, we had a species that was not a helpless victim of what nature threw at it but one that could be proactive in meeting new challenges. Other primates had shown a limited ability in this direction, in creating primitive tools to obtain food — a stone to crack a nut or a stick to insert into a termite nest to tease out the tasty insects, for example. However, some creatures with much more limited brainpower, i.e. some birds, had equally ingenious stratagems for obtaining food — such as dropping cockles from a height to break their shells. Man's ability to combat nature's challenges was of a completely different order. If the temperature fell then he could create fire and use the skins of animals as clothing to keep himself warm enough to survive. He could hunt animals much larger and stronger than himself by the use of stone-tipped weapons. He had the imagination to devise hunting strategies to defeat even the largest prey; by the use of fire, mammoths could be panicked into plunging over a precipice to their death. In the very earliest of man's activities, we can see the beginning of his ability to master the forces of nature to a limited extent.

Although early man showed these extraordinary abilities, they were not at a level that interfered with the general balance of Nature. Early mankind had a tough fight to survive. Infantile mortality was high and, in the dangerous activity of hunting, fatalities were frequent. To some larger carnivores of the times, such as sabre-toothed tigers, man would have been the prey rather than the predator. Any wound would be likely to lead to gangrene and death; indeed life would have been brutal and short. In the circumstances only a tiny proportion of the population would have survived to late adulthood, say to an age of forty years, and the great majority that survived infancy would have died much sooner. However, from the point of view of the survival of the species, a limited adult lifespan would not have mattered over-much. At the crudest level the main function of any species is just to propagate itself and once an organism passes through the stage of greatest fertility then it becomes surplus to requirements, a useless consumer of resources. As an example of this principle in operation, once a female salmon has laid her eggs she usually dies; she is of no more use to the species of which she is a member. At a more extreme level, once a male black-widow spider has fertilised his mate then it is customary for the female to eat her ex-lover; not only is he surplus to requirements but he becomes an asset as part of the food resource of the species.

Mankind passed through various stages — from hunters to hunter-gatherers and then to farmers — but in all these activities man was, initially, just a part of Nature. This state of affairs, where man applied his brainpower to the struggle for survival but did not seriously interfere with the balance of Nature, lasted until the first civilisations arose some 8,000 years ago. The characteristic of a civilisation is that it is a complex form of society with central control and a specialisation of activities. In ancient Egypt, the population was divided into various categories. There were the peasants, who worked on the land and kept the whole population fed. Mostly living in cities, there were artisans of various kinds, who produced and sold the artefacts that society needed. It was felt necessary to protect this society from outsiders, or to subjugate outsiders for the profit of the state, so a military structure existed to engage in both offensive and defensive operations. The priesthood, charged with the duty of understanding and predicting the inundations of the Nile, studied the heavens as an aid to defining time and in doing so they became the first scientists. At the top of this structure were the rulers, the Pharaoh and the court, the individuals who made the big decisions that controlled the overall activity of the state. Religion, which had become part of the belief of even primitive men, had become formalised with the priesthood as the intermediaries between the population at large and the gods. The Egyptians believed in eternal life, and mummification of the bodies of the Pharaohs and other important individuals and the provision of goods for the next life led, firstly, to the understanding of chemical processes for mummification and, secondly, to the need for building huge structures, the pyramids. To build a pyramid required an enormous mobilisation of labour and it is likely that much of this labour came from the peasantry at quiet times in the agricultural calendar. An American space scientist, when asked about the proportion of the US national effort going into space research replied that, 'If the US put the same proportion of its national effort into space research as the Egyptians put into building pyramids then the US would have been able to put a pyramid into orbit'.

What we see in these early civilisations is not so much interference with Nature but, through large scale activity, the first signs that man might be able to do so. The Great Wall of China did not change the Earth's environment but it *can* be seen from space. Nevertheless, despite these early indications, for several millennia after the first civilisations arose there was little influence of man on Nature, and technological advance was steady but slow. A Chinese Rip van Winkle who fell asleep in the Xia Dynasty in 2000 BCE would have found little to astonish him if he awoke in 15th Century Europe.

38.3 The Rules of the Game Change — Man Modifies the Environment

The last 500 years have been very different, with the rapid and accelerating advance of technology. There was the introduction of explosives into warfare, and the increased building of ships, for both commercial and military use, had a significant effect on the tree population in the British Isles. Advances in medicine and medical procedures began to have an effect on the average lifespan. The big change, which altered in a fundamental way the interaction of man and Nature, was the *Industrial Revolution*, which began in Britain at the end of the 18th Century. A great deal of manual labour was replaced by machines driven by some power source or other. Some of the power came from water wheels but much of it came from steam engines driven by coal, and the mining of coal grew apace with the increase of demand. Carbon, which had existed as carbon dioxide and methane in the primitive Earth's atmosphere, had been transformed into vegetation which then became buried and, under high pressure, turned into almost pure carbon, safely locked away in the bowels of the Earth. Now man was digging it out and restoring it to the atmosphere whence it originally came. Not a good idea on the face of it! Other changes were also taking place. Medicine was being improved to the point where both the length and quality of life were being greatly improved. Having a bad gene no longer necessarily meant an early death and the pos-sibility of passing on bad genes greatly increased. The survival of the unfit now became a new reality that operated in parallel with Darwinian evolution.

The increases in productivity in both food and industrial production enabled the population of the world to greatly increase. In 1650 the human population of the world had been about 600 million. By 1750 it had risen to 800 million, an increase of 33% in 100 years. In 1850 the world population had reached 1,200 million, an increase of 50% in 100 years. The world population in 1950 was 2,200 million, an increase of more than 80% in 100 years. In the year 2000 the population was 6,000 million, an increase of 170% in 50 years. It is predicted that the rate of increase of population will slow down so that in 2050 there will be 9,200 million people, an increase of *only* 53% in 50 years. This increase in population puts enormous strains on the world's resources. To meet an almost insatiable demand for hardwood, equatorial rainforests are being destroyed at a great rate in places such as Brazil and some parts of East Asia. Trees are part of the process that removes carbon dioxide, a greenhouse gas, from the atmosphere and the fewer trees there are the less is the amount of carbon dioxide removed.

The strain on the Earth and its resources does not come equally from all the individuals forming the teeming billions of the Earth's population. A high proportion of those living today are living at a subsistence level, barely finding enough to eat and owning very few possessions. Many starve. At the other end of the scale are the world's most affluent societies where the burning problems are where to go for the next vacation, which car to buy to replace the present one and the appalling lack of choice in the local boutique. If the poorest people in the world were to achieve even 20% of the consumption of the most affluent, the ability of the Earth to provide the necessary resources would be strained to breaking point.

The general principle that seems to be accepted by many societies is that 'life is good so more life must be better'. The idea that an exponential increase in population would lead to a crisis because of limited resources was first put forward by Thomas Malthus (Figure 38.1), a British economist. He forecast that the population could grow much faster than the food supply so that eventually the world would not be able to feed itself. The terms in which Malthus saw the problem may not be the terms in which we see it today — but a problem there is.

Figure 38.1 Thomas Malthus (1766–1834).

Mankind has now reached the point where it is adversely affecting the very environment in which it lives. If these changes were occurring as a consequence of natural non-catastrophic events then perhaps there would be time for adaptation to take place but there is no way that a species can react to changes taking place on a timescale of decades. To give an example of the way that man affects the environment, consider the changes in the Aral Sea, a body of water between Kazakhstan and Uzbekistan that was previously under the control of the Soviet Union. In the 1930s, two rivers that fed into the Aral Sea were diverted to provide water to irrigate a desert region which could then be used to grow cotton and various foodstuffs. The plan worked in its primary objective but the Aral Sea has dried up, has split into two parts, and occupies about one quarter of its previous area (Figure 38.2). The exposed seabed is covered with salt and toxic chemicals which have been spread by the wind to neighbouring areas thus making them polluted and sterile. Attempts are being made to improve the situation but these have been only partially successful.

The greatest potential threat to the Earth today is that of global warming. The science of the greenhouse effect is sound and testable

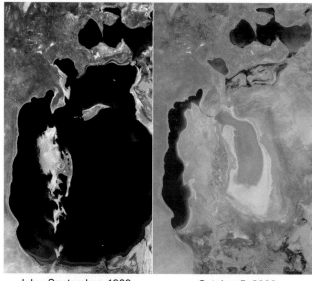

July – September, 1989   October 5, 2008

Figure 38.2   Aerial photographs of the Aral Sea (NASA).

but there are many complications in the way that the warming process could work. For example, there is the possibility that a slightly warmer climate will give more cloud cover to the Earth, so increasing its albedo and reducing solar heating, which would naturally providing a check on too great an increase of temperature. Another factor, in the other direction, is that a great deal of methane, a highly effective greenhouse gas, is permanently stored in permafrost regions, such as large swathes of Siberia, and heating will melt some of the permafrost and so accelerate the release of greenhouse gasses. Indeed, this effect has already been observed. We cannot be *absolutely* sure, but the balance of evidence is that global warming *will* occur and that it will be harmful to human life as a whole, although there may be some regions of the world where it could actually be beneficial to life. There is even a slight possibility that, because of interactions with the environment that we have not anticipated (remember the Aral Sea), global warming could spiral out of control to make human life impossible anywhere on Earth.

The precautionary principle tells us that we should treat this problem seriously but it requires a political consensus, both nationally and internationally, that is difficult to achieve. Developing nations argue, with some justification, that it is the developed nations that have caused most of the problem and have gained the most benefit from their actions, so they should bear the brunt of applying the remedies. Again, while most people would claim that they would make any sacrifice for their children and grandchildren, a politician who asks them to reduce their standard of living by 10% for the benefit of their grandchildren is unlikely to be elected.

Karl Marx stated in *Das Kapital* that capitalism contained the seeds of its own destruction but what he claimed applied to capitalism probably applies in a wider sense. For example, many of the world's most beautiful places have attracted tourists, whose physical needs have been met by building concrete and glass hotels that, by their very nature, detract from the beauty that led to their construction. Perhaps mankind has evolved with too large a brain — which will become the agency of its own destruction. Mankind, as a species, could have existed indefinitely with bows and arrows, windmills and the Black Death, but it may not be able to survive very long with nuclear weapons, coal-fired power stations and antibiotics.

Chapter 39

Musing Again

I am idly staring out of my study window again and I spot our local squirrel — we call him Claude. He has a drink from the bowl of water put there for the birds and then smartly shins up the nearby tree. Since a local restaurant recently introduced squirrel into its menu, my wife and I are always a little anxious if we do not see Claude for two or three days. He is a grey squirrel and, by and large, they get a poor press. Our British native red squirrels are now confined to a few isolated regions of the country where they are free of competition from their grey North American cousins. Do we have nationalism with squirrels, I wonder? After all, the grey squirrels are better able to compete under British conditions so why should Darwin's evolutionary theory not apply to squirrels? Still, I do admit, the red squirrels look cute and I wish them well.

The local squirrels seem to be doing well at present especially as the autumn has been mild and has provided an abundance of food. Last year we had a very cold winter, with continuous snow cover for several weeks although, prior to that, we had not had bad snow conditions in this part of Yorkshire for many years — a sign of global warming perhaps. Another hard winter is forecast for this year — we hope that Claude will survive and that we shall again have the pleasure of seeing him nimbly scampering up the silver birches that grace our neighbourhood.

I wonder what squirrels think about, assuming that they think at all. They must do, I suppose, when they collect nuts and store them in their secret larders. On the other hand I think a great deal — as René Descartes said, '*cogito ergo sum*', 'I think, therefore I am', so I certainly *am*. How did the first life begin? What was the transition that turned chemistry into life? Is there any sense in the question 'What triggered the Big Bang?' Are there questions to which there are no answers? Are there questions to which there *are* answers but which require a greater intelligence than ours to comprehend? Can religion provide answers or is it just sweeping problems under the carpet and substituting one huge insoluble problem for a number of lesser insoluble problems? I am put in mind of the television adverts that urge you to replace your many small debts by one large consolidated debt.

Does the fact that I think about science, politics and philosophy make me superior to the squirrel? Indeed, is there any objective measure of better or worse in judging the success or the quality of one species over another? I doubt it really. Perhaps the only objective measure of success in an evolutionary sense is survivability and, on that basis, the cockroach and the shark come out well. They have been around for a long time — about 400 million years. It is also claimed that, if man ever created a nuclear hell-on-Earth, the best, and perhaps only, survivors on land would be cockroaches. They would carry forward the banner of future evolution, perhaps culminating in a few hundred million years in another kind of intelligent creature, possibly one that knew how to survive with its intelligence.

I wonder what the world will be like forty generations from now — will it be a wondrous place full of marvels that would delight us to the same extent that our technology would delight a citizen of ancient Rome? Alternatively, will it be a wasteland supporting only life forms that can exist in ambient temperatures of 50°C? I wonder!

Index